parasites of freshwater fishes

a review of their control and treatment

Dr. Glenn L. Hoffman[1] and Dr. Fred P. Meyer[2]
Bureau of Sport Fisheries and Wildlife,
Division of Fishery Research
edited by Dr. John C. Landolt,
Shepherd College, Shepherdstown, W. Va.

[1]Parasitologist, Eastern Fish Disease Laboratory,
Leetown (P.O. Kearneysville), West Virginia 25430

[2]Director, Fish Control Laboratory, La Crosse, Wisconsin 54601

Front cover—Apistogramma wickleri on nest; eggs being destroyed by fungus. Photo by H. J. Richter
Back cover—Apistogramma wickleri on nest; eggs healthy. Photo by H. J. Richter

This book is dedicated to the many federal and state fish hatchery biologists (fish pathobiologists) who have given us encouragement and help over the past fifteen years.

Glenn L. Hoffman
Fred P. Meyer

Distributed in the U.S.A. by T.F.H. Publications, Inc., 211 West Sylvania Avenue, P.O. Box 27, Neptune City, N.J. 07753; in England by T.F.H. (Gt. Britain) Ltd., 13 Nutley Lane, Reigate, Surrey; in Canada by Clarke, Irwin & Company, Clarwin House, 791 St. Clair Avenue West, Toronto 10, Ontario; in Southeast Asia by Y. W. Ong, 9 Lorong 36 Geylang, Singapore 14; in Australia and the south Pacific by Pet Imports Pty. Ltd., P.O. Box 149, Brookvale 2100, N.S.W., Australia.

CONTENTS

TREATMENT INDEX

Names in parentheses are indexed synonyms

INTRODUCTION

The recent increased interest in all aspects of fish culture has created a need for information on the control of fish parasites. Data and references on fish-parasite control efforts are widely scattered and many of the articles are in obscure journals. This review attempts to bring together many of the references related to this subject. It is hoped that it will serve as a reliable source concerning research efforts published largely through 1970 and that it will provide a stimulus for work in new areas and on new concepts of parasite control, such as the search for systemic parasiticides (R. Allison, 1969).

No thorough attempt is made to evaluate the various works, although the success or failure of the treatments (as reported by the authors) is indicated for the various compounds. Persons considering the use of any of the reported techniques are advised that it should first be tested on a small lot of fish before it is used on a large population. The degree of success encountered in the use of any compound is related to many factors. Workers testing compounds should be aware that water chemistry, temperature, pH, salinity, the presence of interfering substances, and formulation differences may reduce or enhance the activity of chemicals. In addition, the susceptibility of fish to toxicants is affected by the developmental stage, age, size, and sometimes even the sex of the fish. Species differences may be profound for both the fish and the target organisms, and the worker should make every effort to determine the species of both. The physiological condition of the fish also influences the success of a treatment. Debilitated or heavily parasitized fish may be unable to tolerate otherwise safe and effective treatments.

The factors contributing to epizootics of fish parasites, both in nature and in culture, have not been thoroughly studied. Because parasites must be considered in relationship to their environment, the following facts may be involved:

1. **Presence of the parasite.** There may be low-grade infestation of an existing fish population, which increases when conditions are favorable, or the parasite may be introduced into a "clean" population.
2. **Presence of the appropriate fish host species.** Some parasites require a specific host, others may infect many species of fish.
3. **Susceptibility of host fish.** Very young and/or weakened fish are often more susceptible than healthy ones. Sometimes partial immunity is developed after an initial sublethal exposure.
4. **Density of parasites.** There must be enough to assure contact with the fish.
5. **Temperature.** Both fish and parasites respond to their optimum temperature with rapid growth. Epizootics are likely when the optimum temperature for the parasite is unfavorable for the host.
6. **Oxygen.** Fish may be debilitated in a low-oxygen environment. Some parasites have demonstrated optimum oxygen levels.
7. **Water current.** Some parasites prefer flowing water, others need quiet water.
8. **pH.** Both parasites and hosts demonstrate optimum pH requirements; *e.g.*, *Costia* disappears from fish below pH 5.3.
9. **Organic matter.** An enriched medium contributing to growth of protozoa may also encourage growth of parasites or commensals which feed on these organisms.
10. **Other fauna and flora.** Intermediate hosts are required by some parasites, and the flora may influence the intermediate hosts.

Many parasites which cause serious disease in fish hatcheries are normally present in low numbers in natural habitats. Under conditions of adequate food and favorable environment for the fish, disease is seldom a threat to the survival of the species. Normally, the parasite feeds on "surplus" tissues and does not endanger the host. Epizootics occur when some factor disrupts the balance between parasite and host—*e.g.*, starvation, pollution, and over-population which debilitate the host, or excessive numbers of parasites, overwhelming the host and causing its death and the death of the parasite as well. Observers thus should be alert to identify the underlying causes of disease.

Some control of parasitic diseases has been achieved in natural waters but the major part of this book is aimed at control of parasitic diseases of cultured or captive fishes.

For comprehensive discussions of fish disease treatment and control methods, the reader is referred to Agapova (1966), Amlacher (1961b), Bauer (1959, 1966), Bauer and Uspenskaya (1959), Davis (1953), Grabda (1965), Hoffman (1970), Ivasik, Karpenko and Sutyagin (1967), Leitritz (1959), Markevich (1967), Musselius (1967), Pavlovskii (1959), Reichenbach-Klinke (1966), Schäperclaus (1954), Tesarcik and Havelka (1967), or Van Duijn (1956, 1967).

To determine the pathogen involved in a particular problem, the reader will find Hoffman (1967) helpful. Other useful works include Davis (1953), Haderlie (1953), Hoffman and Sindermann (1962), Hugghins (1959), M. Meyer (1962), Northcote (1957), Reichenbach-Klinke and Elkan (1965), Sindermann (1953), Van Cleave and Mueller (1934), and Wellborn and Rogers (1966). European parasites can be identified with the aid of Bykhovskaya-Pavlovskaya, *et al.* (1962).

Tables of conversion units and guides for the determination of quantities of chemicals needed to yield desired concentrations may be found on pages 17–21.

Information for determining the identity and properties of the various compounds may be found in The Merck Index (1968), Pesticide Index (Frear, 1961), Farm Chemicals 1970 Handbook (Berg 1970), The American Drug Index (Wilson and Jones, 1962), and Gardner and Cooke (1968). Toxicity information may be found in McKee and Wolf (1963) and pages 169–186.

Any fish to be used for human or animal food should be treated with drugs and chemicals only in accordance with current laws and regulations. Federal agencies having such regulations are the Food and Drug Administration and the Environmental Protection Agency; local and State agencies may also have regulations.

At the present time, according to Food and Drug Administration regulations, Sulfamerazine and Terramycin for certain bacterial diseases may be added to the feed of fish intended for human consumption. Rotenone and antimycin may be used for fish control and eradication. TFM (3-trifluormethyl-4-nitrophenol) and a combination of TFM with Bayer 73 are registered for use to control the parasitic sea lamprey in the Great Lakes waters. Copper sulfate may be used as an algicide. Bayer 73 (Bayluscide) is registered for use as a molluscicide only in Puerto Rico and Michigan; the 5% granular form sinks and reportedly does not kill swimming fish. MS-222 (tricaine methanesulfonate) has been approved as a fish anesthetic. Toxic chemicals, not approved by the Food and Drug Administration or Environmental Protection Agency, must not be added to water reaching human or agricultural water supplies.

The authors acknowledge the assistance of the U.S. Fish and Wildlife Service and thank the many U.S. and State Fish Hatchery Biologists (Pathobiologists) who provided much information and encouragement during this project.

TREATMENT METHODS

Techniques for the treatment of parasitized fishes must be adapted to each individual situation. Factors such as the value of the infected fish, virulence of the parasite, cost of treatment, chance of success, likelihood of spread of the disease, and the condition of the fish all must be considered in choosing the technique best suited to the occasion.

While quarantine cannot be considered a treatment, it is a useful tool in the control of parasitic diseases. Suspect fish should be held in isolation from other stocks until it can be determined that they are free of parasites. If diseased fish cannot be treated effectively, they should be destroyed to prevent spread of the disease beyond the immediate area.

Various methods of treatment are employed by fish culturists. In the tables listing specific treatments, they will be identified by letter. The following types of treatment are in current use:

Biological (B)—In nature, certain organisms prey upon parasites or upon intermediate hosts required for a parasite to complete its life cycle. Examples of such relationships include the redear sunfish *(Lepomis microlophus)* which feeds on snails; the protozoan, *Amphileptus voracus*, which feeds on other parasitic protozoans; and *Chaetogaster*, an oligochaete worm, which feeds on snails. While such relationships may be employed as preventive or prophylactic measures, at present they are seldom effective in the control of an epizootic. Whenever possible, however, biological control methods should be used. For further information, see specific items in the text.

Dip (D)—Dip treatments are employed whenever fish or eggs can be readily handled and can be returned to fresh, parasite-free water immediately after treatment. Since high concentrations of chemical are used frequently for strictly specified, short periods of time, leaving the fish or eggs in the solution beyond the recommended time would be lethal to the fish. Amounts of chemical required for dip treatments are low but labor costs are usually high.

Flush (F)—Flush treatments are employed in two ways. Usually, the chemical is added to a tank and left for a specified period of time after which the solution is drained from the tank and replaced with fresh water. In the other technique, the chemical is added to the incoming water supply continuously for a specified period of time and the solution is allowed to flow through the tank or raceway. After the desired time has elapsed, fresh water is again introduced. This technique is not applicable to pond situations.

Indefinite (I)—In treatments of this type, chemical is added to the water and left indefinitely. Degradation or dissipation of the chemical occurs within the pond or aquarium with no dilution by fresh water. No handling of the fish is required but the cost of treatment may be high due to the amount of chemical required. This type of treatment is most commonly used in pond situations and occasionally in aquariums.

Management (M)—Occasionally a parasite problem can be controlled by techniques of managing the fish populations. Examples of this type of technique might be disinfection of facilities; draining ponds and thoroughly drying them just prior to stocking fish to reduce snail populations; removal of roosting and nesting trees that attract fish-eating birds which are carriers of helminth parasites; or the eradication of other fish that might serve as intermediate hosts for parasites. Keeping susceptible stages of fish in concrete tanks until they are less susceptible to a parasitic disease is another way in which this technique could be applied.

Oral (O)—If fish are feeding and will accept artificial feeds, it is possible to add medications to the diet. Anthelmintics, antiprotozoan compounds, and antibiotics may be administered in this manner.

Physical (P)—Occasionally alteration of the environmental conditions can be achieved to the extent that a particular parasite is no longer able to develop. Raising the water temperature, increasing the flow rate, filtration, sonic vibrations, altering the pH, or ultraviolet light disinfection are examples of this type of treatment. Bauer (1959) discusses these matters more fully. Wherever feasible, physical treatment methods should be employed in preference to chemical methods.

Surgical (S)—Surgical methods have drastic limitations. In isolated instances in which the fish involved are particularly valuable (as in the case of broodfish), it may be feasible to remove attached parasites manually with forceps or by surgery.

Topical (T)—Valuable fishes such as ornamentals or broodfish are sometimes treated by applying chemical directly to the affected areas. Lesions or attached parasites may be treated by the application of concentrated chemical.

TREATMENT CONVERSION CHART

(Amounts listed are for active ingredients or a trade name preparation, depending on the recommendations)

parts per million (ppm)	mg per liter	mg per gal	oz (avoir) per 1,000 gal	g per cu ft	oz per 1,000 cu ft	lbs per acre ft
0.1	0.1	0.38	0.013	0.0028	0.1	0.27
1	1	3.8	0.134	0.0283	1	2.7
2	2	7.6	0.268	0.0567	2	5.4
3	3	11.4	0.402	0.0851	3	8.1
4	4	15.2	0.536	0.1134	3.99	10.8
5	5	19.0	0.670	0.1418	4.99	13.5
6	6	22.8	0.804	0.1701	5.99	16.2
7	7	26.6	0.938	0.1985	6.99	18.9
8	8	30.4	1.072	0.2268	7.99	21.6
9	9	34.1	1.206	0.2552	8.98	24.3
10	10	38.0	1.340	0.2835	9.98	27.0
11	11	41.8	1.474	0.3118	10.98	29.7
12	12	45.6	1.608	0.3401	11.98	32.4
13	13	49.4	1.742	0.3684	12.97	35.1
14	14	53.2	1.876	0.3967	13.97	37.8
15	15	57.0	2.010	0.4250	14.98	40.5
16	16	60.8	2.144	0.4533	15.97	43.2
17	17	64.6	2.278	0.4816	16.97	45.9
18	18	68.4	2.412	0.5099	17.96	48.6
19	19	72.2	2.546	0.5382	18.96	51.3
20	20	76.0	2.680	0.5620	19.97	54.0
100	100	380.0	13.40	2.8350	99.84	270.0
1,000	1,000	3,800.0	134.0	28.3500	998.4	2,700.0

CONVERSION TABLE

Acre (A)	= 43,560 square feet
	a square 208.71 feet per side
	a circle with a diameter of 235.4 feet
	0.405 hectare
Acre-foot	= 1 acre of surface covered with 1 foot of water
	43,560 cubic feet
	2,718144 pounds of water
	325,850 gallons
	1,233,342 liters
Centner	= 1 Zentner
	50 kilograms
Cubic Centimeter (cc)	= See Milliliter
Cubic Foot (ft^3)	= 1,728 cubic inches
	7.48 gallons
	62.4 pounds of water
	28.32 liters
Cubic Meter (m^3)	= 1,000 liters
	a cube one meter per side
CWT	= 100 pounds
	45.3 kilograms
Doppel Zentner (dz)	= 100 kilograms
Dram (dr)	= 1.772 grams
Foot (ft)	= 12 inches
	30.48 centimeters
	0.305 meter
Gallon (gal) (U.S.)	= 8.34 pounds of water
	3,785 milliliters
	3.785 liters
	0.1337 cubic feet
	4 quarts (U.S.)
Gallon (Imperial)	= 4.8 U.S. quarts
Gallons per minute	(gpm)
Grain (gr)	= 64.8 milligrams
	0.065 grams
	0.35 ounce

Gram (g)	= 1,000 milligrams
	0.001 kilograms
	0.0353 ounce
	0.0022 pound
Hectare	= 2.47 acres
	10,000 square meters
Hundredweight (cwt)	= 100 pounds
	45.3592 kilograms
Kilogram (kg)	= 1,000 grams
	2.205 pounds
Kilohertz (kHz)	= 1,000 cycles per second
Kiloliter (kl)	= 1,000 liters
	264.18 gallons
	35.315 cubic feet
Kilometer (kilom or km)	= 0.62 miles
Liquid ounce	= 29.57 milliliters (do not use when preparing solutions or concentrations)
Liter (l)	= 1,000 milliliters
	1 kilogram of water
	35.28 ounces
	1.057 quart
	2.2 pounds
	0.035 cubic foot of water
Meter (m)	= 39.37 inches
	3.28 feet
	100 centimeters
Microgram (mcg, μg)	= 0.001 milligram
Microwatt second (mws)	= Microwatt seconds/cm^2
Mile (mi)	= 1.61 kilometers
Milligram (mg)	= 1,000 mcg
	0.001 grams
	0.0154 grain
Milliliter (ml)	= 0.001 liter
	1 gram of water
	0.002 pound of water
	0.0003 gallon (U.S.)
Minutes (min)	

One percent in food	= 4.5 grams per pound of feed
	0.2 ounces per pound of feed
	83 grains per pound of feed
	10,000 parts per million in feed
One percent solution	= 1 gram in 100 milliliters
	38 grams in one gallon (U.S.)
	1.3 ounces in one gallon (U.S.)
	1 ounce in 0.75 gallon (U.S.)
	4.53 grams in one pound
	0.624 pounds in one cubic foot
Ounce (oz)	= 28.35 grams (See also liquid ounce)
Pint (pt)	= 473.2 milliliters
Pound (lb)	= 16 ounces
	453.6 grams
	453.6 milliliters of water
	7,000 grains
	0.12 gallon of water
	0.016 cubic foot of water
One part per million (ppm)	= 1 milligram per liter of water
	1 gram per 264 gallons of water
	0.0038 gram per gallon of water
	0.0283 gram per cubic foot of water
	1 ounce per 1,000 cubic feet of water
	2.7 pounds per acre foot of water
	1 gram per cubic meter of water
	1 pound in 999,999 pounds of water
Quart (qt)	= 946.36 milliliters
	0.95 liter
	0.25 gallon
	2.086 pounds of water
	0.0334 cubic foot
Square centimeter (cm^2 or sq. cm.)	
Square foot (ft^2)	= a square 12 inches per side
	930 square centimeters
Square meter (m^2)	= a square 100 centimeters per side
	10,000 square centimeters
	10.76 square feet
Stone	= 14 pounds

Tablespoon (tbsp)	= 15 milliliters 3 teaspoons
Teaspoon (tsp)	= 5 milliliters
Ton (Metric)	= 1,000 kilograms 2,204.6 pounds
Ton (T) U.S.	= 2,000 pounds 906 kilograms
Yard (yd)	= 3 feet 36 inches 91.44 centimeters 0.914 meter
Zentner	= See Centner

TREATMENT CONVERSION CHART FOR FORMALIN (37-40% COMMERCIAL FORMALDEHYDE)[1]

parts per million (ppm)	equivalent in thousands	ml per liter	ml per gal	pints per 1,000 cu ft	quarts per acre ft
1	1:1,000,000	0.001	0.0038	0.0598	1.304[2]
2	1:500,000	0.002	0.0076	0.1196	2.608
3	1:333,333	0.003	0.0114	0.1794	3.912
4	1:250,000	0.004	0.0152	0.2392	5.216
5	1:200,000	0.005	0.0190	0.2990	6.520
6	1:166,666	0.006	0.0228	0.3588	7.824
7	1:142,857	0.007	0.0266	0.4186	9.128
8	1:125,000	0.008	0.0304	0.4784	10.432
9	1:111,111	0.009	0.0343	0.5382	11.736
10	1:100,000	0.010	0.0380	0.5980	13.04
15	1:66,666	0.015	0.057	0.897	19.56
20	1:50,000	0.020	0.076	1.196	26.08
25	1:40,000	0.025	0.095	1.495	32.60
100	1:10,000	0.100	0.38	5.98	130.40
166	1:6,000	0.166	0.63	9.93	216.46
200	1:5,000	0.200	0.76	11.96	260.80
250	1:4,000	0.250	0.95	14.95	323.00
1,000	1:1,000	1.000	3.80	59.80	1304.00

[1]ml = milliliter (cubic centimeter), gal = gallon. Local druggist can probably supply a dropper calibrated to deliver milliliters. For measuring small amounts one can determine the number of drops a dropper will deliver into one milliliter, *e.g.*, many eye droppers will deliver approximately 20 drops of water per milliliter.
[2]One gallon of formalin will yield 3 ppm in one acre-foot of water.

A SYNOPSIS OF THE MOST WIDELY USED THERAPY AND CONTROL OF FISH PARASITES

This synopsis represents those techniques used most widely in treating fish for parasitemia. Inclusion in this listing in no way implies FDA or USDA approval for their use. Caution should always be exercised when using a chemical for the first time or when treating a species or life stage whose tolerances are unknown. For further information, including references, see appropriate section in text.

I. External fungi *(Saprolegnia, Achlya, Aphanomyces, Leptomitus,* and *Pythium)*

A. On fish eggs

(1) Malachite green—5 ppm as a one hour flush used daily (trout)

(2) Malachite green—1,500 ppm as a 10 second dip (catfish)

(3) Formalin—2,000 ppm for 15 minutes (trout)

B. On fish

(1) Malachite green—66 ppm for 10 to 30 second dip (trout)

(2) Malachite green—0.1 ppm for 1 hour flush (warmwater fishes—catfish, bass, *etc.*)

(3) Malachite green—1 to 3 ppm for 1 hour flush (trout—use with caution)

II. External Protozoans

A. *Ichthyophthirius* (ichthyophthiriasis, "Ich"):

Chemical treatment for this parasite must be used daily or the chemical must be in contact until "Ich" disappears (5 to 10 days). If all fish are heavily infected, usually no known treatment will save them. It is best to examine fish regularly and start treatment at the first finding of "Ich." Quarantine of suspect fish is recommended.

(1) 25 ppm of mixture of malachite green and formalin (14 gms of malachite green per 1 gallon formalin)—

use for up to 6 hours daily in tanks or raceways. Use at 3 to 4 day intervals in ponds.

(2) Formalin—

(a) 250 ppm for 1 hour daily if water temperature 10°C or less;
200 ppm for 1 hour daily if water temperature is 10 to 15°C;
166 ppm for 1 hour daily if water temperature is above 15°C.

(b) 15 to 25 ppm in ponds or aquariums on alternate days until control is achieved. Oxygen depletion may follow application in hot weather.

(3) Malachite green (oxalate)—

(a) 0.1 to 0.15 ppm at 3 to 4 day intervals in ponds or aquariums.

(b) 2 ppm daily for 30 minutes (trout).

(4) Copper sulfate

(a) Less than 1 ppm if calcium carbonate level in water is less than 50 ppm (use with extreme caution).

(b) 1 to 2 ppm if calcium carbonate level is 50 to 200 ppm.

(c) 2 ppm mixed with 3 ppm citric acid is sometimes used in ponds with calcium carbonate levels above 200 ppm.

(5) Temperature—for tropical fish that can resist high temperatures, raise the temperature to 90°F for 6 hours daily for 3 to 5 days.

B. Other external protozoa (*Costia, Chilodonella, Oodinium,* trichodinids, *Ambiphrya, Apiosoma, Epistylis, Trichophrya*):

The methods for ichthyophthiriasis are usually satisfactory; one application ordinarily suffices.

III. Internal Protozoans

A. *Hexamita (Octomitus)*:

Although at least some strains are not pathogenic, some fish culturists believe *Hexamita* to be harmful.

(1) Dimetridazole— 0.15% in food daily for 3 days.

(2) Enheptin—0.2% in food for 3 days.

(3) Cyzine—20 ppm in dry food for 3 days.

(4) Carbarsone oxide—0.2% in food for 3 days, but arsenicals must not be used for fish intended for human or livestock food.

B. *Myxosoma cerebralis* (whirling disease), *Ceratomyxa shasta,* and probably other Myxosporida:

(1) Drugs—there is little information on practical treatment with systemic drugs. Stovarsol and furazolidone[1] show promise.

(2) Do not transfer infected fish to waters that do not contain these organisms.

(3) Destroy all infected fish, disinfect the hatchery, and make certain that the water supply does not contain infected fish. The best known pond disinfectant is quicklime, 1 ton to the acre of drained, but moist pond bottom.

The European method for the control of *M. cerebralis,* rearing trout in spore-free water until 7 cm long, and then transferring them to contaminated waters, is dangerous because it creates asymptomatic disease carriers.

C. *Eimeria* (coccidiosis):

(1) Drugs—Stovarsol at 1 mg/g of food, and Furoxone, show promise.

(2) Disinfection of drained ponds—calcium hypochlorite is used in Russia at 500 kg/hectare (446 lbs. to the acre).

IV. Monogenetic Trematodes (*Gyrodactylus* and dactylogyrids)

A. Formalin—use as for ichthyophthiriasis; one application usually suffices.

B. Dylox (Dipterex, Neguvon, Chlorophos, Trichlorofon, Foschlor)—0.25 ppm in ponds and aquariums.

C. Bromex-50—0.18 ppm in ponds.

D. Potassium permanganate—five to 10 ppm for 1 to 2 hours; 3 to 5 ppm in ponds and aquariums.

[1]R. E. L. Taylor, Animal Science Dept., University of Nevada, Reno 89507 (personal communication).

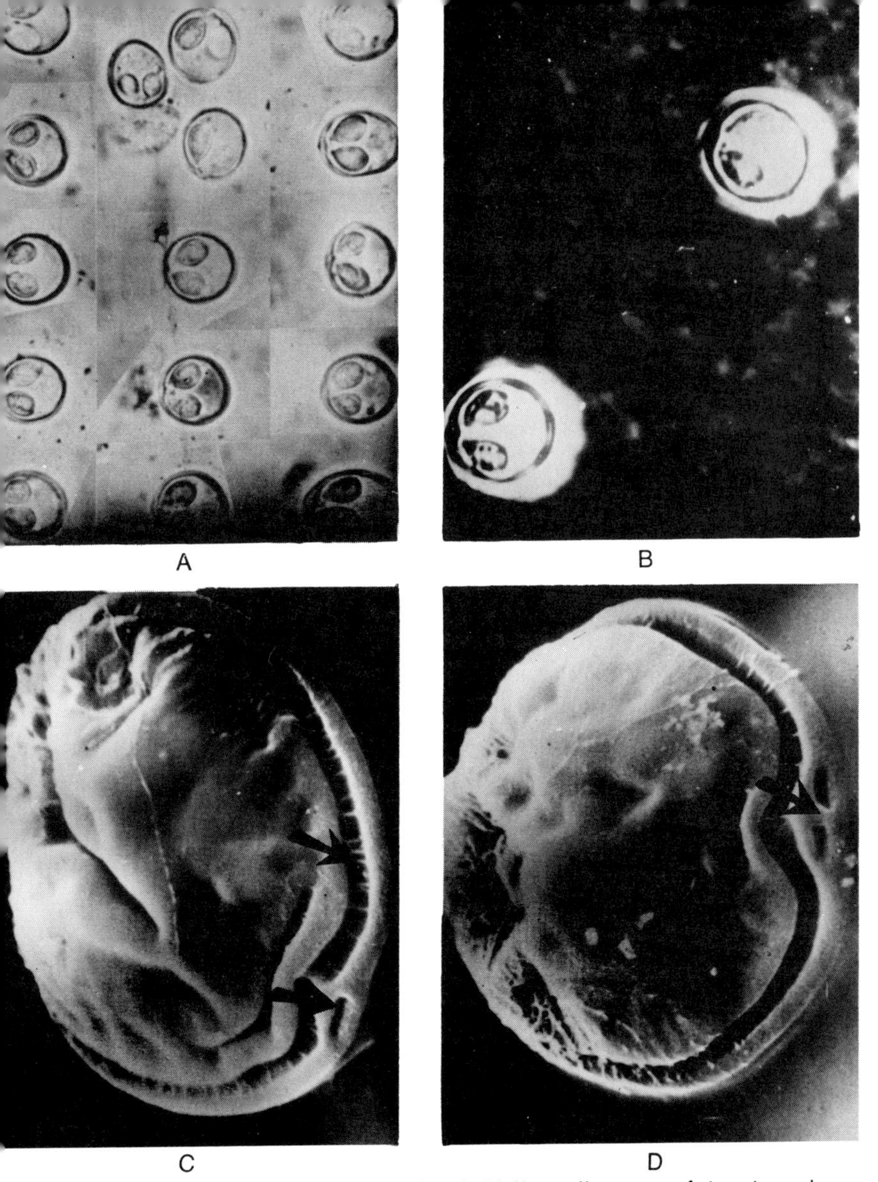

Spores of *Myxosoma cerebralis* (whirling disease of trout and salmon). From Lom and Hoffman (1971); courtesy of Journal of Parisitology.

(a) Variation in shape and size. × 3000.

(b) Mucoid envelope (halo), India ink preparation. × 3000.

(c) Spore in upper anterior view showing spectacular groove and polar filament pore, scanning electron photomicrograph. × 11,000.

(d) As (c) but the pore seems to be split into two.

V. Digenetic Trematodes (flukes)

A. Di-n-butyl tin oxide or dibutyltin dilaurate—250 mg/kg of fish or 0.3% of food for 5 days.

B. Filtration of water supplies—removes cercariae of fish trematodes.

C. Snails (intermediate hosts of trematodes)—see copper sulfate and Frescon under molluscicides.

VI. Cestodes (tapeworms)

A. Di-n-butyl tin oxide—250 mg/kg of fish or 0.3% of food for 5 days.

B. Dibutyltin dilaurate—used as in (A) is probably effective.

C. Fish population management—the incidence of *Triaenophorus* larvae was reduced by increasing fishing pressure on the adult host, *Esox*.

D. Food control—bass tapeworm can be controlled by eliminating infected forage fish from the diet. *Eubothrium* and *Diphyllobothrium* of trout were controlled by removing infected copepods from the water supply by sand-gravel filtration and treating the fish with (A) or (B).

VII. Nematodes (roundworms)

A. Santonin—0.04 g/fish, has been used for sturgeon.

VIII. Acanthocephala

A. Di-n-butyl tin oxide—shows promise, see cestodes.

B. Bithionol—used in France at 0.2 g/kg of fish, mixed at 2% in feed for convenience.

IX. Hirudinea (leeches)

A. Dylox (Dipterex, Neguvon, Chlorophos, Trichlorofon, Foschlor)—0.5 to 1 ppm.

X. Parasitic Copepods

The following methods are satisfactory for killing the larval stages.

A. Dylox (Dipterex, Neguvon, Chlorophos, Trichlorofon, Foschlor)—0.25 to 0.5 ppm 5 times at weekly intervals; 1 ppm when temperature is above 29°C. This apparently kills adult *Argulus* and larval *Lernaea*, but not adult *Lernaea*.

B. Bromex-50—0.12 ppm in ponds at weekly intervals.

C. Copepods can be removed with forceps from individual valuable fish.

XI. Glochidia

No chemical treatment known. Remove clam larvae from water supply with sand filters or wire screens, 200 mesh.

XII. Snail control

A. Management—thoroughly drain and dry ponds at least once per year to keep snail populations at a minimum.

B. Physical removal—good pond sanitation (removal of detritus by shovel or high pressure fire hose) sometimes reduces snail populations.

C. Copper sulfate and copper carbonate—

(1) If methyl orange alkalinity is less than 50 ppm use 2 lbs of copper carbonate to 1,000 ft^2 of pond bottom or copper sulfate at less than 1 ppm, with great caution.

(2) If methyl orange alkalinity is 50 ppm or greater, use 2 lbs copper sulfate plus 1 lb copper carbonate to 1,000 ft^2 of pond bottom, or 1 ppm copper sulfate in the water.

D. Bayluscide—0.5 to 1 ppm in water, will control all stages of snails but is toxic to fish. Fish may be restocked 3 weeks after treatment. The granular (sinking) formulation can be used in the presence of fish that won't eat the granules; it may be toxic to bottom-dwelling fish.

E. Frescon—0.01 to 0.1 ppm, kills snails but not fish; is being used in Great Britain and Africa; use cautiously.

XIII. Miscellaneous

A. Predaceous insects

(1) Floating oils such as kerosene, diesel fuel, motor oil, and vegetable oil—2.5 to 5 gal to the acre kill air-breathing beetle larvae.

(2) Dylox—0.5 ppm kills predaceous aquatic insects.

B. Crayfish

(1) Baytex—0.23 ppm kills crayfish but not fish.

C. Small crustacea (intermediate hosts of helminths)

(1) Chlorophos (Dylox)—0.8 ppm is effective in Europe.

(2) Calcium hypochlorite (HTH)—probably effective for most parasites, but is toxic for fish. Use with caution.

GLOSSARY

acanthocephala — spiny-headed worms
algae — chlorophyll-bearing, single cell or filamentous plants
algicide — chemical that kills algae
anthelmintic — drug used to combat helminths
antibiotic — compound derived from one living organism which will kill or suppress other organisms
antifungal — chemical or process that kills or inhibits fungi
antimicrobial — antagonistic to microbes
antiseptic — compounds or procedures to avoid dirt and kill disease organisms
asymptomatic — inapparent infection
bacteriostat — chemical that inhibits bacteria
cathartic — purgative; induces evacuation of intestine
cercaria — free-swimming larval stage of digenetic trematode
cestodes — flatworms, usually segmented, long, and with specialized "head" (scolex)
coccidiostat — drug that inhibits coccidea
copepodid — larval stage of copepod
copepods — small crustacea, sometimes called water fleas
Cyclops — genus of small copepod
debilitated — weakened
digenetic trematodes — trematodes having intermediate hosts
disinfectant — chemical or procedure to kill specific disease organisms
ecto- — external
enzootic — a retention of low incidence of a disease in an animal population
epizootic — a rise in incidence and intensity of a disease in an animal population
fecundity — reproductive ability
fish hatchery biologist — see pathobiologist
formulation — mixture of active ingredients and carrying materials to make a commercial product

fungi	— filamentous or spore-forming plants without chlorophyl
fungicide	— fungus-killing chemical
herbicide	— chemical that kills plants
infection	— internal invasion
infestation	— external attack
insecticide	— chemical that kills insects
in vitro	— handled in glass or plastic containers
malariastat	— drug that inhibits malaria
miticide	— chemical that kills mites
molluscicide	— chemical that kills molluscs
molluscs	— group of animals that include snails and clams
monogenetic trematode	— trematodes that have no intermediate hosts and are usually ectoparasitic
nanoplankton	— microscopic small aquatic plants and animals
nauplius	— first larval stage of copepod
nematodes	— cylindrical unsegmented round worms
ppm	— parts per million
parasiticide	— chemical that kills parasites
pathobiologist	— one who diagnoses and studies fish diseases and recommends or suggests measures for their control
protozoa	— single-celled, small animals; may possess flagella, cilia, pseudopods or be predominantly spores
protozoicide	— chemical that kills protozoa
schistosomiasis	— disease caused by *Schistosoma* (blood fluke of mammals) also called "Bilharzia"
scoliosis	— lateral spinal curvature
sterilize	— procedure that kills all living organisms
systemic drugs	— drugs which circulate throughout the body
tomite	— small motile daughter cell of ciliated protozoa
trematodes	— parasitic flatworms without segments but with suckers (flukes)
trophozoite (troph)	— active, growing stage of protozoa
Tubifex	— oligochaete annelid, common in mud with high organic content
zoospore	— motile free-living stage of fungus

TREATMENT LISTS BY PARASITE GROUPS

ORGANIZATION OF TABLES

Treatment tables are provided for each of the major parasite groups. The various genera within each group are arranged alphabetically in column one, using the same format as found in Hoffman (1967). The control of the sea lamprey, *Petromyzon marinus*, has not been included; the latest review is that of Schnick (1972). Treatments[1] used on the various parasite genera are listed alphabetically in column three.[2] Chemical synonyms may be found on p. 187. Hosts on which the parasites were found are listed in column two. Although *in vitro* is not a host, it is included in the host column—by *in vitro* we mean tests done on isolated parasite organisms. Some of the "unidentified" parasites and hosts are our omissions, not the cited researchers; time did not permit correcting these and we thought it wise to include them, even if incomplete.

[1]Mention of product names in this book does not imply endorsement by the authors or the U.S. Fish and Wildlife Service.

[2]In Method column: B = biological, D = dip, F = flush, I = indefinite length of time, M = management of fish population, draining ponds, etc., O = Oral, P = physical, S = surgical, T = topical (see p. 15 for greater detail).

Table 1. FUNGI ON FISH EGGS

Parasite	Host	Treatment	Dosage	Method	Number of Applications	Frequency	Author's Report of Success	Remarks	References
Saprolegnia sp.	*Esox* eggs	Globucid (sodium)	200 ppm	F 8 hrs.	?	?	Not effective	....	Schäperclaus, 1954
,, ,,	?	Halamid	10 ppm	I	1	?	....	Slowed growth	Deufel, 1970
,, ,,	Salmon eggs	Malachite green	5 ppm	F 1 hr.	?	?	Effective	....	Johnson, Adams & McElrath, 1955
Unidentified Fungus	Fish eggs	Copper sulfate	1 oz of saturated solution	F	?	Daily	Effective	....	Schneberger, 1941
,, ,,	Fish eggs	Copper sulfate	5 ppm	F 1 hr.	?	....	Effective	....	Hoffman, 1969
,, ,,	Salmon eggs	Formalin	5,000 ppm	F 30 min.	?	Weekly	Effective	Dosage excessive	Watanabe, 1940
,, ,,	Salmon eggs	Formalin	2,000 ppm	F 15 min.	?	Weekly	Effective	....	Burrows, 1949
,, ,,	Trout eggs	Formalin	1,666 ppm	F 15 min.	?	As needed	Effective	....	Reddecliff, 1958 and Steffens, 1962
,, ,,	Trout eggs	Malachite green	500 ppm	F	?	As needed	Effective	Do not use within 24 hours of hatching	O'Donnell 1947
,, ,,	Trout eggs	Malachite green	5 ppm	F 1 hr	?	As needed	Effective	May cause crippling	Burrows, 1949; Bradford, 1966
,, ,,	*Stizostedion* eggs	Malachite green	6 g to each battery	F	?	5 day intervals	Effective	Flow was 50 gpm	Cummins, 1954
,, ,,	*Salvelinus namaycush* eggs	Malachite green	90 ml of 1% soln to head of trough	F	?	Twice daily	Effective	Flow was 6 gpm	Robertson, 1954
,, ,,	*Esox lucius* eggs	Malachite green	10 ppm	F 15 min.	?	Alternate days	Effective	....	Gottwald, 1961

,,	,,	*In vitro*	Malachite green	1–5 ppm	F 1 hr.	?	?	Effective	Some species of fungi inhibited by 1 ppm for 5 min.	Martin, 1968
,,	,,	Sturgeon eggs	Malachite green	5 ppm	F 30 min.	?	Daily	Effective	....	Astakhova and Martino 1968
,,	,,	*Ictalurus punctatus* eggs	Malachite green	1,500 ppm	D 15 sec.	4	Daily	Effective	Do not use within 24 hours of hatching	F. Meyer, Unpubl.
,,	,	*Cyprinus carpio* eggs	Malachite green	10 ppm	F	1	?	Effective	Increased hatch by 25%	Askerov 1968
,,	,,	*Lepomis macrochirus* eggs	Malachite green	0.005 ppm	I	?	Twice daily	Effective	Change water twice each day	Merriner, 1969
,,	,,	*Lepomis gibbosus* eggs	Malachite green	0.01 ppm	I	?	Twice daily	Effective	Change water twice each day	Merriner, 1969
,,	,,	Trout eggs	Ozone	26–65 ppm	I	?	Continuous	Inhibited growth	Recirculation system, no controls in test	Benoit and Matlin, 1966
,,	,,	Fish eggs	Rivanol	10 ppm	F	?	?	Not effective	....	Schäperclaus, 1954
,,	,,	Fish eggs	Trypaflavine	10 ppm	F	?	?	Not effective	....	Schäperclaus, 1954
,,	,,	Sturgeon eggs	Ultraviolet light	3 BUV–60 P or 3 BUV–30 P bactericidal lamps 3.9 w	I	?	Continuous	Effective	8 m^3 per hour flow. Applied to water supply only	Astakhova and Martino 1968
,	,,	*In vitro*	Ultraviolet light	35,000 MWS	I	?	Continuous	Effective	Kills zoopores; fish not included	Vlasenko, 1969
,,	,,	Sturgeon eggs	Ultraviolet light	MBU–3 PRK–7 lamp, 35 w	I	?	Continuous	Effective	Flow was 4.5 gpm; water supply treated	Kokhanskaya, 1970

Fungus growing on fish eggs. Photo by Frickhinger.

Table 2. FUNGI ON FISH

Parasite	Host	Treatment	Dosage	Method	Number of Appli-cations	Frequency	Author's Report of Success	Remarks	References
Achlya sp.	In vitro	Malachite Green	5–15 ppm	F 30 min–1 hr.	1	....	Inhibitory	....	Martin, 1968
Branchiomyces sp.	*Cyprinus carpio* & *Esox lucius*	Calcium cyanamide	Not given	M	1	?	Assumed effective	Used to disinfect tanks	Amlacher, 1961b
,, ,,	*Micropterus salmoides* & *Roccus saxatilis*	Formalin	25 ppm	I	2	Alternate days	Inhibitory	Did not eradicate disease	F. Meyer, Unpubl.
,, ,,	*Cyprinus carpio* & *Esox lucius*	Quicklime	Not given	M	?	?	Assumed effective	Used to disinfect tanks	Amlacher, 1961b
Ichthyophonus hoferi	Aquarium fish	Chloramphenicol	0.001 ppm in diet	O	....	....	Unknown	....	Van Duijn, 1967
,, ,,	Trout ponds	Chlorine	200 ppm	M	....	....	Assumed effective	Used to disinfect pond	Erickson, 1965
,, ,,	Trout ponds	Malachite green	290 ppm	M	....	....	Assumed effective	Used to disinfect pond	Erickson, 1965
,, ,,	Unspecified	Phenoxetol	100–200 ppm	?	?	....	....	....	Reichenbach-Klinke, 1966
Saprolegnia sp.	*Carassius auratus*	Acriflavine	3 ppm	I	1	....	Effective	In aquarium	Yousuf-Ali, 1968
,, ,,	*Notemigonus crysoleucas*	Acriflavine	10 ppm	I	1	....	Not effective	....	F. Meyer, Unpubl.
,, ,,	*Notemigonus crysoleucas*	Chlortetracycline	10 ppm	I	1	....	Not effective	....	F. Meyer, Unpubl.
,, ,,	Unidentified	Collargol	0.1 ppm	F 20 min.	1	....	Effective	....	Amlacher, 1961b; Reichenbach-Klinke, 1966
,, ,,	Aquarium fish	Collargol	1–3 ppm	F 1–3 hrs.	1	....	Effective	Not as effective as malachite green	Van Duijn, 1967

,,	,,	*Micropterus* sp.	Copper sulfate	1 ppm	?	?	....	Not effective	....	Mellen, 1928
,,	,,	Unidentified	Copper sulfate	500 ppm	D 1 min.	?	As needed	Not effective	....	Davis, 1953
,,	,,	Unidentified	Copper sulfate	100 ppm	F 10–30 min.	?	?	Not effective	....	Amlacher, 1961b
,,	,,	*Notemigonus crysoleucas*	Copper sulfate	1.0 ppm	I	1	....	Not effective	....	F. Meyer, Unpubl.
,,	,,	*Notemigonus crysoleucas*	Cyprex	4 ppm	I	1	....	Inhibitory	....	F. Meyer, Unpubl.
,,	,,	*Notemigonus crysoleucas*	Ferbam	8 ppm	I	1	....	Not effective	....	F. Meyer, Unpubl.
,,	,,	*Notemigonus crysoleucas*	Formalin	25 ppm	I	1	....	Inhibitory	....	F. Meyer, Unpubl.
,,	,,	*Notemigonus crysoleucas*	Gentian violet	0.3 ppm	I	1	....	Effective	Toxic to weak fish	F. Meyer, Unpubl.
,,	,,	Unidentified	Griseofulvin	10 ppm	I	1	....	Effective	Should be tested experimentally	Reichenbach-Klinke, 1966
,,	,,	*In vitro*	Furanace	10 ppm	I	1	....	Effective	....	Shimizu and Takase, 1967
,,	,,	*Notemigonus crysoleucas*	Iodoform	2 ppm	I	1	....	Inhibitory	Not curative	F. Meyer, Unpubl.
,,	,,	*Notemigonus crysoleucas*	Karathane	0.1 ppm	I	1	....	....	Toxic to fish	F. Meyer, Unpubl.
,,	,,	Unidentified	Malachite green	66 ppm	D 10–30 sec.	?	Daily	Effective	....	Foster and Woodbury, 1936
,,	,,	*Micropterus* sp.	Malachite green	66 ppm	D 30 sec.	?	As needed	Effective	....	O'Donnell, 1941
,,	,,	*Notemigonus crysoleucas*	Malachite green	0.2 ppm	I	1	....	Effective	Toxic to many species of fish	F. Meyer, Unpubl.

[*continued*

Saprolegniasis, *Scatophagus argus.* Photo by Frickhinger.

External fungus (large white patches), *Micropterus dolomieui*, smallmouth bass. Photo by G. Hoffman.

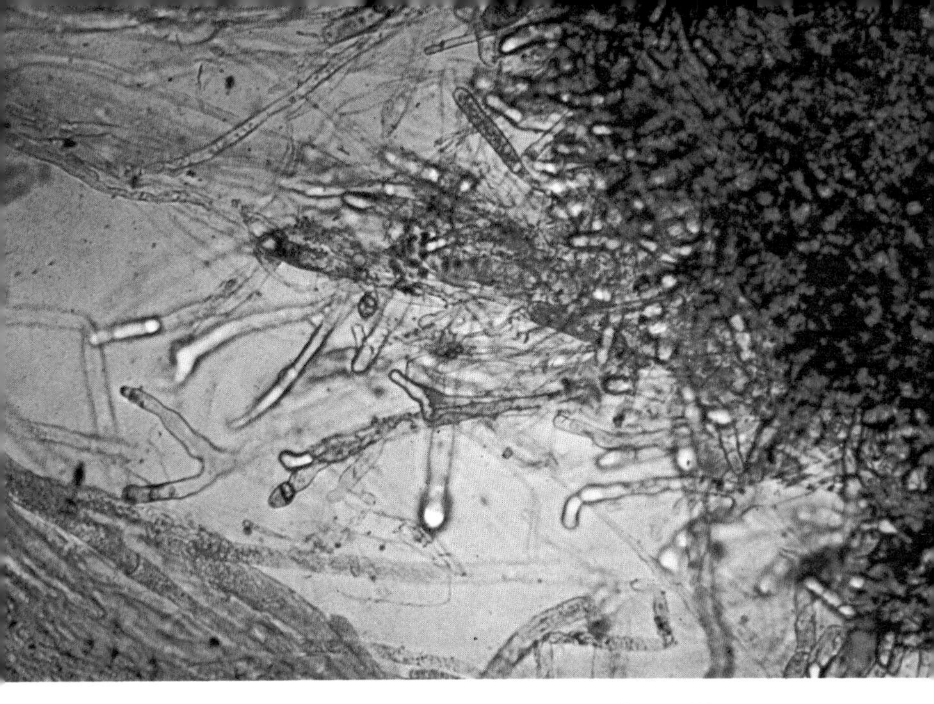

Saprolegnia, microscopic wet mount. Courtesy USF & WS.

Branchiomycosis (gill rot) of the gills of carp. Note eroded areas. Courtesy of P. de Kinkelin, Laboratoire d'Ichthyopathologie, Route de Thiverval, 78 Thiverval-Grignon, France.

Table 2. Fungi on Fish—*continued*

Parasite	Host	Treatment	Dosage	Method	Number of Applications	Frequency	Author's Report of Success	Remarks	References
Saprolegnia sp	*In vitro*	Malachite green	2.0 ppm	F 1 hr.	?	?	Effective	Toxic to fish	Scott and Warren, 1964
,, ,,	*In vitro*	Malachite green	1 ppm	F 30 min–1 hr.	1	....	Inhibitory	....	Martin, 1968
,, ,,	*Notemigonus crysoleucas*	Methylene blue	10 ppm	I	1	....	Not effective	....	F. Meyer, Unpubl.
,, ,,	Unidentified	Mercurochrome	10 ppm	F 12 hrs.	1	....	Effective	....	Seale, 1928
,, ,,	Salamander embryos	Mercurochrome	1.34 ppm	?	?	....	Effective	....	Detwiler and McKennon, 1929
,, ,,	*Notemigonus crysoleucas*	Oxytetracycline	10 ppm	I	1	....	Not effective	....	F. Meyer, Unpubl.
,, ,,	Aquarium fish	Penicillin	1 ppm	I	?	Daily	?	Could induce resistant strains of bacteria	Van Duijn, 1967
,, ,,	Aquarium fish	Phenoxetol	100 ppm	F 12 hrs.	?	?	?	....	Rankin, 1952
,, ,,	Aquarium fish	Phenoxetol	100 ppm	I	1	....	Effective	Not toxic to plants	Loader, 1963
,, ,,	*In vitro*	Phenoxetol	5,000 ppm	I	?	?	Effective	Toxic to fish	Scott and Warren, 1964
,, ,,	*Notemigonus crysoleucas*	Potassium dichromate	20 ppm	I	1	....	Not effective	....	F. Meyer, Unpubl.
,, ,,	Pond fishes	Potassium permanganate	10 ppm	F $1\frac{1}{2}$ hrs.	?	?	Effective	....	Plehn (in Schäperclaus, 1954)
,, ,,	Unidentified	Potassium permanganate	10 ppm	F 30 min.	?	?	Effective	....	Amlacher, 1961b; Reichenbach-Klinke, 1966

,,	,,	*In vitro*	Potassium permanganate	5,000 ppm	I	?	?	Effective	Toxic to many fish	Scott and Warren, 1964
,,	,,	Unidentified	Silver nitrate	10,000 ppm	T	?	As needed	Effective	....	Reichenbach-Klinke, 1966
,,	,,	Unidentified	Sodium chloride	30,000 ppm	D until fish show distress	?	As needed	Effective	....	Davis, 1953
,,	,,	*Notemigonus crysoleucas*	Sodium chloride	20 ppm	I	1	....	Not effective	....	F. Meyer, Unpubl.
,,	,,	*Notemigonus crysoleucas*	Thiram	1.0 ppm	I	1	....	Not effective	....	F. Meyer, Unpubl.
,,	,,	Unidentified	Ultraviolet light	10,000 MWS/cm²	P	....	Continuous	Effective	Min. effective dose for hyphae	Normandreau, 1968
,,	,,	Unidentified	Ultraviolet light	39,564 MWS/cm²	P	?	Continuous	Effective	Min. effective dose for zoospores	Normandreau, 1968
,,	,,	Unidentified	Ultraviolet light	35,000 MWS	I-P	?	Continuous	Effective	Kills zoospores in water supply	Vlasenko, 1969
,,	,,	*In vitro*	Ultraviolet light	MBU-3(PRK-7)	P	?	Continuous	Effective	Dosage not given	Kokhanskaya, 1970
,,	,,	*Notemigonus crysoleucas*	Ziram	2 ppm	I	1	....	Inhibitory	Not curative	F. Meyer, Unpubl.
,,	,,	*Notemigonus crysoleucas*	Ziram	4 ppm	I	1	....	Inhibitory	Toxic to fish	F. Meyer, Unpubl.
,,	,,	Plants	....	....	....	....	....	....	A review of fungicides used on plants	Bent, 1969
Unidentified fungus		Unidentified	Aquarol	80 ppm	D	3	3 day interval	Effective	....	Amlacher, 1961b; Reichenbach-Klinke, 1966

[*continued*

Table 2. Fungi on Fish—*continued*

Parasite	Host	Treatment	Dosage	Method	Number of Applications	Frequency	Author's Report of Success	Remarks	References
Unidentified fungus	Salmonids	Diquat	2–4 ppm	?	?	?	Inhibitory	....	J. Wood, 1968
,, ,,	Tropical fish	Griseofulvin	500 mg/tank	I	6	Daily	Effective	No experiments cited	Sandler, 1966
,, ,,	Pond fish	Malachite green	67 ppm	D 10–30 sec.	?	As needed	Effective	....	R. Allison, 1954
,, ,,	Pond fish	Malachite green	5.5 ppm	F 45 min.	?	As needed	Effective	Used in ponds	R. Allison, 1954
,, ,,	Salmonids	Malachite green	3.3 ppm	F 1 hr.	1	....	Effective	....	Fish, 1939
,, ,,	Salmonids	Malachite green	1.25 ppm	F $1\frac{1}{2}$ hrs.	1	....	Effective	....	Fish, 1939
,, ,,	Pond fish	Malachite green	0.1 ppm	I	?	?	Effective	Used in ponds	R. Allison, 1954
,, ,,	Unidentified	Malachite green	0.02 ppm	I	?	?	Inhibitory	....	Arasaki, Nozawa, and Mizaki, 1958
,, ,,	*Ictalurus punctatus*	Malachite green	1 ppm	F 1hr.	?	As needed	Effective	Toxic to some lots of fish	Clemens and Sneed, 1958
,, ,,	Trout	Malachite green	5 ppm	F 20 min.	?	3 day interval	Effective	....	Sakowicz and Gottwald, 1958
,, ,,	Salmonids	Malachite green	2.5–5 ppm	F 30 min.	1	....	Effective	Used in earthen ponds	Hublou, 1958
,, ,,	*Coregonus* sp.; *Esox* sp.;	Malachite green	10 ppm	F 15 min.	?	Alternate days	Effective	....	Gottwald, 1961
,, ,,	*Ictalurus punctatus*	Malachite green	100 ml of 500 ppm	F	?	?	Effective	In catfish rearing troughs	Snow, 1962
,, ,,	*In vitro*	Malachite green	2 ppm	F 1 hr.	?	?	Effective	Toxic to fish	Scott and Warren, 1964
,, ,,	*Oncorhynchus tschawytscha*	Malachite green	1 ppm	F 1 hr.	1	Twice weekly	Not effective	Toxic to young salmon	Knittel, 1966; J. Wood, 1968
,, ,,	*Oncorhynchus* sp.	Malachite green	100,000 ppm	T	2	Weekly	Effective	....	Knittel, 1966

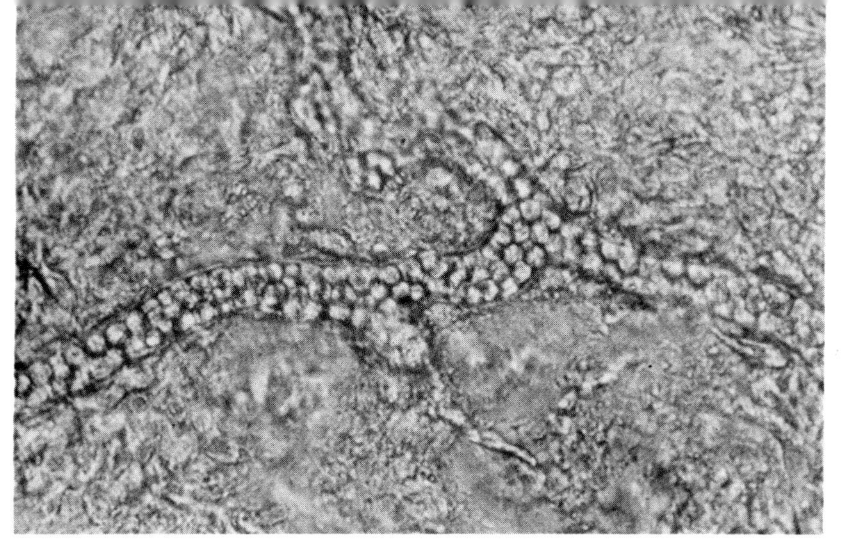

Branchiomyces sp. from gill rot of carp, unstained preparation. Courtesy of Dr. P. Ghittino, Istituto Zooprofilattico Sperimentale del Piemonte e della Liguria, Torino, Italy, and Edizioni Rivista di Zootecnia.

Ichthyophonus hoferi causing bumpy condition and spinal curvature of rainbow trout. Courtesy of D. Erickson, Snake River Trout Company, Buhl, Idaho.

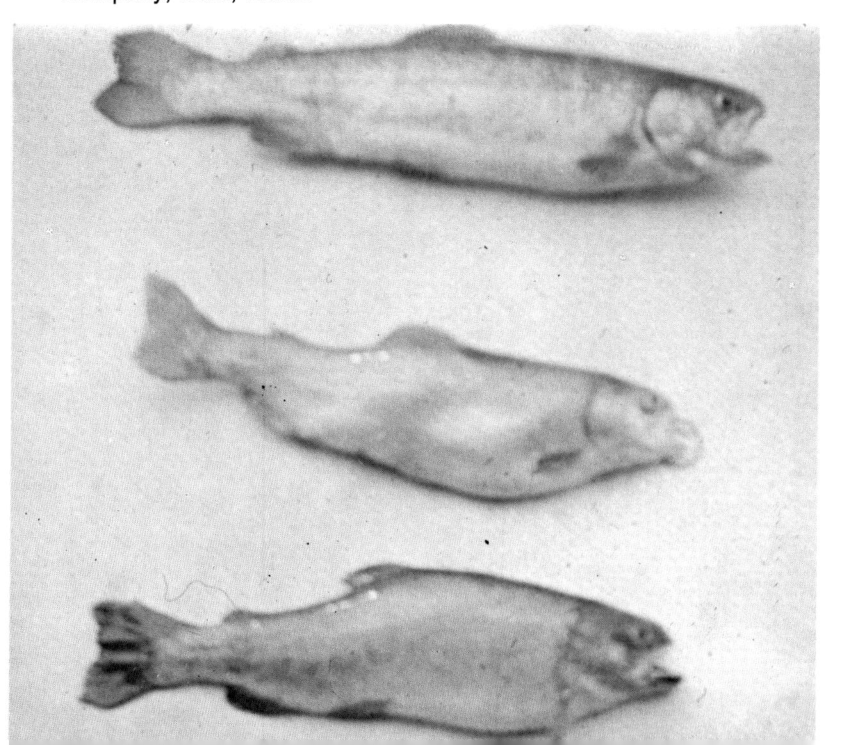

Table 3. PROTOZOANS (EXTERNAL)

Parasite	Host	Treatment	Dosage	Method	Number of Appli-cations	Frequency	Author's Report of Success	Remarks	References
Ambiphrya macropodia	*Ictalurus punctatus*	Formalin	10–15 ppm	I	1	....	Effective	....	Allison, 1963
Ambiphrya pyriformis	Pondfishes	Acetic acid	1,000 ppm	?	1	....	Effective	....	Tripathi, 1954
,, ,,	*Cirrhina reba; Catla*	Formalin	200 ppm	?	1	....	Effective	....	Tripathi, 1954
,, ,,	Pondfishes	Sodium chloride	30,000 ppm	F 5–10 min.	1	....	Effective	....	Tripathi, 1954
Ambiphrya sp. *(Scyphidia)*	*Ictalurus punctatus*	Acriflavine	20 ppm	I	1	....	Inhibitory	....	F. Meyer, Unpubl.
,, ,, ,,	*Notemigonus crysoleucas*	Copper sulfate	0.5–1 ppm	I	1	....	Inhibitory	Do not use in low carbonate waters	F. Meyer, Unpubl.
,, ,, ,,	Trout	Formalin	250 ppm	F 1 hr.	1	....	Effective	....	Davis, 1953
,, ,, ,,	*Ictalurus punctatus*	Iodoform	2 ppm	I	1	....	Not effective	....	F. Meyer, Unpubl.
,, ,, ,,	*Ictalurus punctatus*	Malachite green	0.1 ppm	I	1	....	Effective	....	F. Meyer, Unpubl.
,, ,, ,,	*Ictalurus punctatus*	Methylene blue	50 ppm	I	1	....	Inhibitory	....	F. Meyer, Unpubl.
,, ,, ,,	*Ictalurus punctatus*	Potassium dichromate	20 ppm	I	1	....	Inhibitory	Organism recurs; toxic to many species of fish	F. Meyer, Unpubl.
,, ,, ,,	*Ictalurus punctatus*	Potassium dichromate	2 ppm	I	As needed	Daily	Inhibitory	....	F. Meyer, Unpubl.

,, ,, ,,	*Ictalurus punctatus*	Potassium dichromate	5 ppm	F 16 hrs.	....	As needed	Effective	Toxic to many species of fish	F. Meyer, Unpubl.
,, ,, ,,	Unidentified small fish	Sodium chloride	30,000 ppm	F 5–10 min	As needed	Daily	Effective	....	Tripathi, 1954
,, ,, ,,	*Ictalurus punctatus*	Tag	2 ppm	F 1 hr.	1	....	Effective	....	Clemens and Sneed, 1958
Amoebae (on gills)	Trout, Salmon	Formalin	250 ppm	F 1 hr.	1	....	Effective	....	Anonymous, 1960
Bodomonas rebae	*Cirrhina* sp.; *Catla* sp.	Sodium chloride	30,000 ppm	F 5–10 min.	1	....	Effective	....	Tripathi, 1954
Bodomonas sp.	Unidentified	Sodium chloride	30,000 ppm	F 5–10 min.	As needed	Daily	Effective	....	Tripathi, 1954
Chilodonella cyprini	*Ictalurus punctatus*	Formalin	10–15 ppm	I	1	....	Effective	....	R. Allison, 1969a
Chilodonella sp.	Aquarium fishes	Acriflavine	10 ppm	F 10 hrs.	1	....	Effective	Not toxic to *Lebistes reticulatus*	Reichenbach-Klinke, 1966; Schäperclaus, 1954
,, ,,	Unidentified	Aquarol	80 ppm	I	3	Daily	Effective	....	Amlacher, 1961b; Reichenbach-Klinke, 1966
,, ,,	Unidentified	Aureomycin	0.13 ppm	I	1	....	Effective	....	Amlacher, 1961b
,, ,,	*Cyprinus carpio*	Chloramine-B	20 ppm	I	1	....	Effective	....	Goncharov, 1966
,, ,,	*Carassius vulgaris*	Collargol	2 ppm	I	1	....	Not effective	....	Schäperclaus, 1954
,, ,,	*Ictalurus punctatus*	Formalin	25 ppm	I	1	....	Effective	Do not use during hot weather	F. Meyer, Unpubl.
,, ,,	Unidentified	Globucid	2,000 ppm	F 24 hrs.	1	....	Effective	....	Schäperclaus, 1954
,, ,,	Pond fishes	Lime filter	....	I-M	....	....	Controlled	Raised pH	Bauer, 1959

[*continued*

Ichthyophonus hoferi, massive renal involvement in rainbow trout. Courtesy of Dr. P. Ghittino, Istituto Zooprofilattico Sperimentale del Piemonte e della Liguria, Torino, Italy, and Edizioni Rivista di Zootecnia.

Ichthyophonus hoferi, wet squash from kidney of rainbow trout, showing hypha-like sporulation. Photo by G. Hoffman.

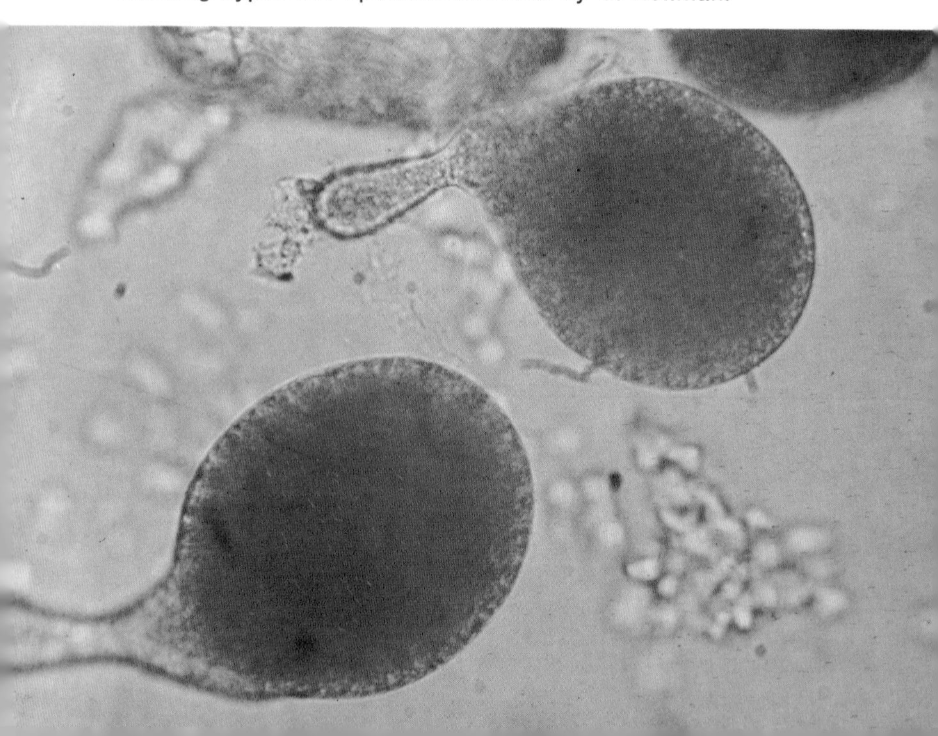

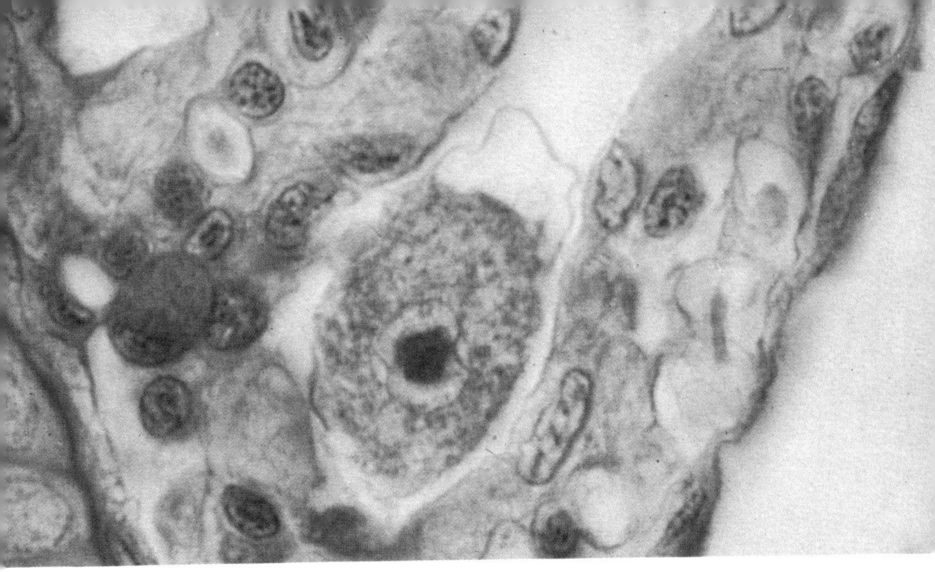

Thecamoeba sp., gill amoeba between gill lamellae of coho salmon yearling, H & E stain, × 400 (T. Sawyer, G. Hoffman, J. Hnath, J. Conrad, Fish Pathology Symposium (1971), AFIP, Univ. Wisc. Press—in press). Photo by Dr. T. Sawyer.

Thecamoeba sp., gill amoeba between gill lamellae of coho salmon yearling, PAS hematoxylin stain × 400 (T. Sawyer, G. Hoffman, J. Hnath, J. Conrad, Fish Pathology Symposium (1971), AFIP, Univ. Wisc. Press—in press). Photo by Dr. T. Sawyer.

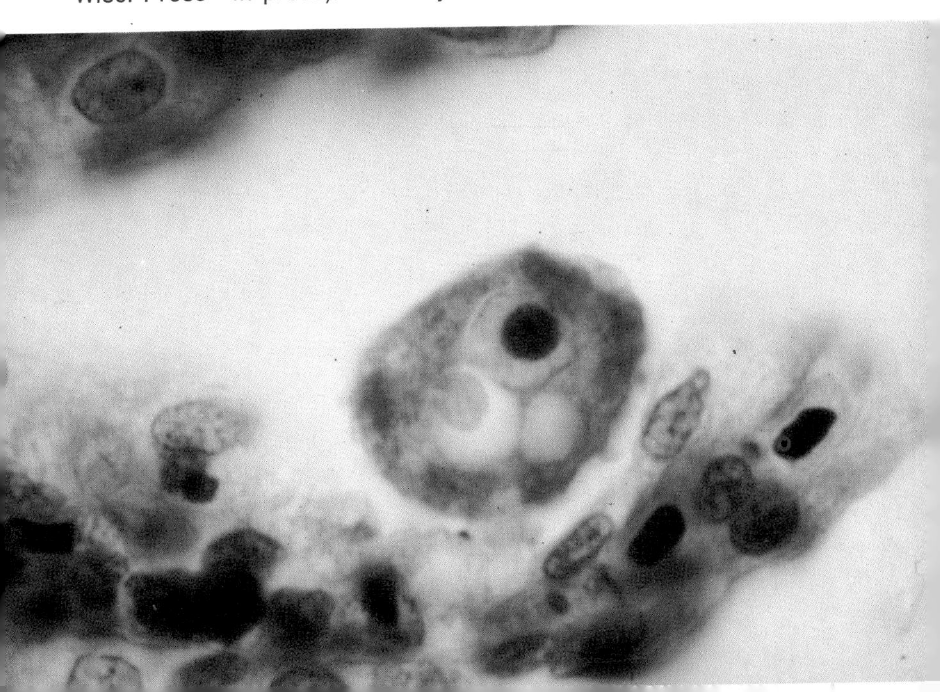

Table 3. Protozoans (External)—*continued*

Parasite	Host	Treatment	Dosage	Method	Number of Applications	Frequency	Author's Report of Success	Remarks	References
Chilodonella sp.	*Cyprinus carpio*	Lysol	200 ppm	D 30 sec.	1	....	Effective	Not as good as formalin	Schäperclaus, 1954
" "	*Cyprinus carpio*, Trout	Malachite green oxalate	0.1–0.15 ppm	I	2–3	Alternate days	Effective	....	Amlacher, 1961a
" "	Unidentified	Methylene blue	3 ppm	I	1	....	Effective	Required several days	Schäperclaus, 1954; Amlacher, 1961b; Reichenbach-Klinke, 1966
" "	*Carassius vulgaris*	Micropur	10 ppm	F 24 hrs.	?	....	Effective	....	Schäperclaus, 1954
" "	Unidentified	Potassium permanganate	10 ppm	F 1 hr.	?	?	Effective	Toxic to *Stizostedion* sp.	Fish, 1933
" "	Unidentified	Potassium permangante	10 ppm	F 90 min.	?	Daily	Effective	....	Amlacher, 1961b
" "	*Cyprinus carpio*	Potassium permanganate	1,000 ppm	D 30–45 sec.	?	Daily	Effective	....	Amlacher, 1961b; Schäperclaus, 1954
" "	Pond fishes	Potassium permanganate	1,000 ppm	F 5-10 min	As needed	Daily	Effective	....	Amlacher, 1961b
" "	Unidentified	Potassium permanganate	1,000 ppm	D 30–40 sec.	As needed	Daily	Effective	....	Amlacher, 1961b; Reichenbach-Klinke, 1966
" "	Unidentified	Potassium permanganate	10 ppm	F 30 min.	As needed	Daily	Effective	....	Reichenbach-Klinke, 1966
" "	*Cyprinus carpio*	Quicklime	2,000 ppm	D 5 sec.	1	....	Effective	....	Schäperclaus, 1954

,,	,,	*Cyprinus carpio* (young)	Quinine hydrochloride	10 ppm	7 hrs.	?	....	Not effective	Toxic to carp fry	Schäperclaus, 1954
,,	,,	*Cyprinus carpio*	Sodium chlorate	50,000 ppm	?	?	?	?	....	Popov and Jankov, 1968
,,	,,	*Cyprinus carpio*	Sodium chloride	10,000 ppm	D 2–10 min.	As needed	Daily	Effective	Avoid zinc containers	Schäperclaus, 1954
,,	,,	*Cyprinus carpio*	Sodium chloride	17,500 ppm	D 3 min.	As needed	Daily	Effective	Avoid zinc containers	Schäperclaus, 1954
,,	,,	*Cyprinus carpio*	Sodium chloride	25,000 ppm	D 20–25 sec.	As needed	Daily	Effective	Avoid zinc containers	Schäperclaus , 1954
,,	,,	*Salmo salar*	Sodium chloride	30,000 ppm	D 5 min.	As needed	Daily	Effective	....	Bauer and Strelkov, 1959
,,	,,	Unidentified small fish	Sodium chloride	15,000 ppm	F 20 min.	As needed	Daily	Effective	Avoid zinc containers	Amlacher, 1961b
,,	,,	Unidentified large fish	Sodium chloride	25,000 ppm	F 10–15 min.	As needed	Daily	Effective	Avoid zinc containers	Amlacher, 1961b
,,	,,	Pond fishes	Sodium chloride	2,000 ppm	I	1	....	Effective	....	Ivasik and Svirepo, 1964
,,	,,	*Cyprinus carpio*	Sodium chloride	4,000 ppm	I	1	....	Effective	....	Bauer, 1966
,,	,,	*Cyprinus carpio*	Temperature	Raise above 10°C	P	1	....	Effective	....	Bauer, 1959
,,	,,	Pond fishes	Temperature	?	P	?	?	?	Epizootic present at 26.7°C	Wellborn, 1966
,,	,,	Unidentified	Trypaflavine	10 ppm	F 2 hrs.	?	?	Effective	....	Schäperclaus, 1954
,,	,,	*In vitro*	Ultraviolet light	1,008,400 MWS/ cm^2	P	1	....	Effective	Fish not included	Vlasenko, 1969
,,	,,	*In vitro*	Ultraviolet light	DRSH-250 for 3 min.	P	1	....	Effective	Fish not included	Laptev, 1967

[*continued*

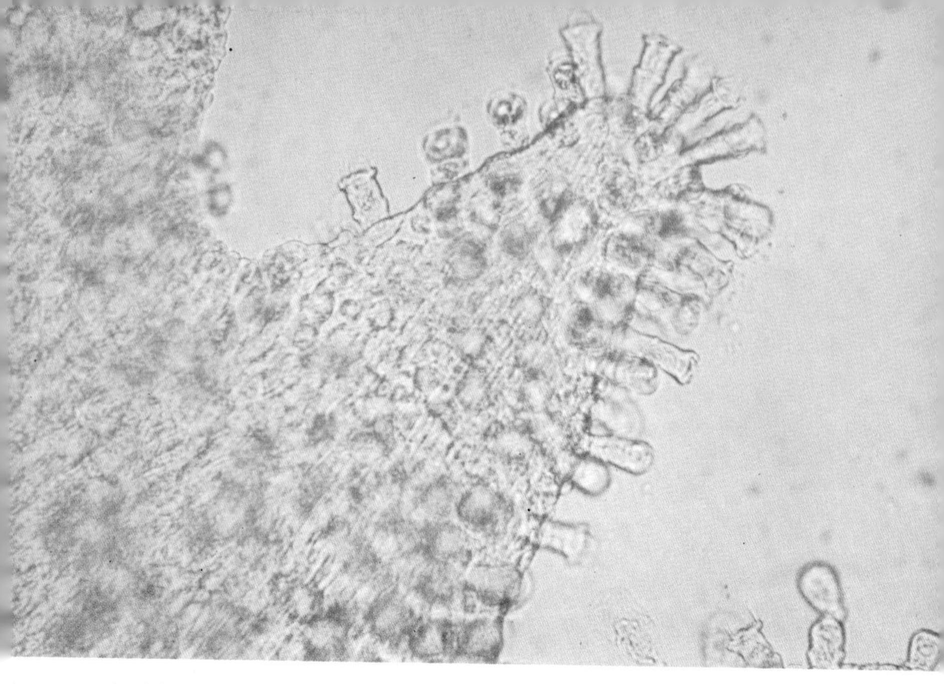

Ambiphrya sp. (*Scyphidia* sp.) on fin of channel catfish. Photo by F. Meyer.

Ambiphrya sp. (*Scyphidia* sp.) on fin of channel catfish. Photo by F. Meyer.

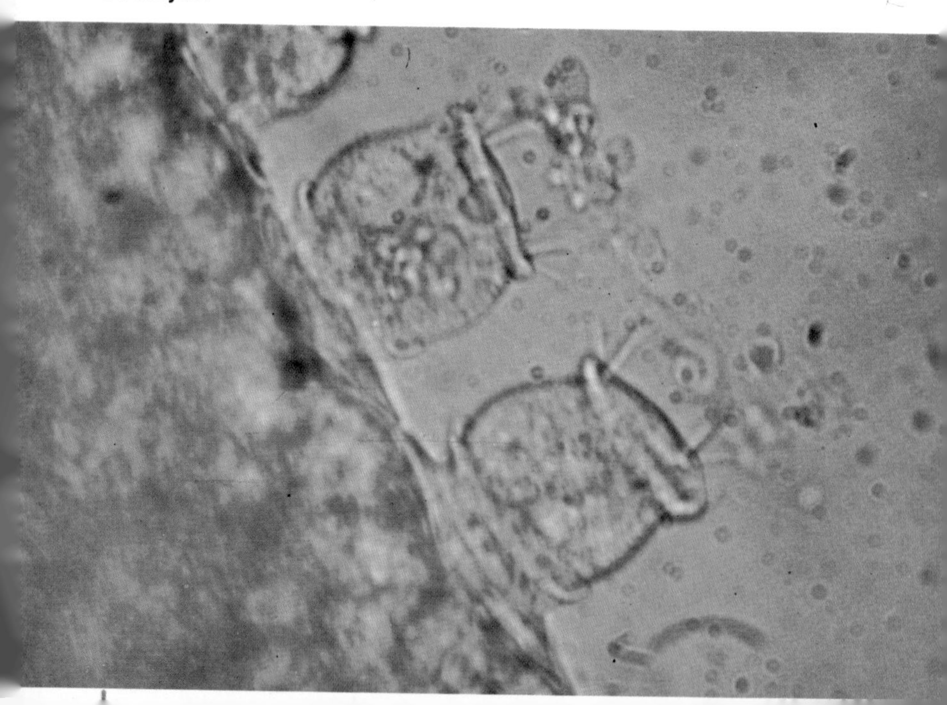

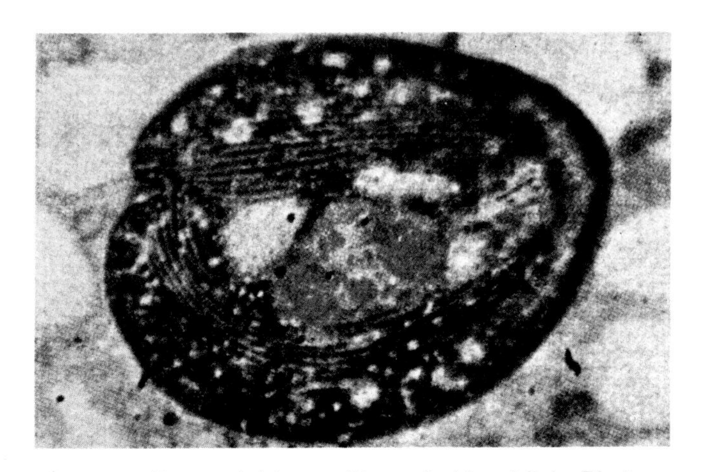

Chilodonella cyprini from gills and skin of fish. Photo by Dr. Jara.

Costiasis (blue slime disease) of rainbow trout. Photo by G. Hoffman.

Table 3. Protozoans (External)—*continued*

Parasite	Host	Treatment	Dosage	Method	Number of Appli-cations	Frequency	Author's Report of Success	Remarks	References
Colponema sp.	*Ictalurus punctatus*	Tag	2 ppm	F 1 hr.	1	....	Effective	....	Clemens and Sneed, 1958
Costia necatrix	Pond fishes	Acetic acid, glacial	2,000 ppm	D	1	....	Effective	....	Hora and Pillay, 1962
,, ,,	Salmonids	Acetic acid, glacial	2,000 ppm	D 1 min. or less	1	....	Effective	....	Davis, 1953
,, ,,	*Ictalurus punctatus*	Formalin	10–15 ppm	I	1	....	Effective	....	Allison, 1963
,, ,,	Pond fishes	Formalin	400 ppm	F 10 min.	1	....	Effective	....	Hora and Pillay, 1962
Costia sp.	Salmonids	Acetic acid, glacial	2,000 ppm	D 1 min.	....	As needed	Effective	....	Davis, 1953
,, ,,	Unidentified	Aquarol	80 ppm	I	3	Daily	Effective	....	Amlacher, 1961b; Reichenbach-Klinke, 1966
,, ,,	Unidentified	Aureomycin	0.13 ppm	I	1	....	Effective	....	Amlacher, 1961b
,, ,,	Unidentified	Chloramin	10 ppm	I	1	....	Effective	Better than 66 ppm	Schäperclaus, 1954
,, ,,	Unidentified	Chloramin	66 ppm	F 2–4 hrs.	1	....	Effective	....	Schäperclaus, 1954
,, ,,	*Cyprinus carpio*	Chloramine-B	20 ppm	I	1	....	Effective	....	Goncharov, 1966
,, ,,	Pond fishes	Copper sulfate	100 ppm	F 10 min.	?	As needed	Effective	....	Amlacher, 1961b
,, ,,	*Carassius auratus*	Copper sulfate	500 ppm	D 1–2 min.	?	As needed	Effective	....	Osborn, 1966
,, ,,	*Carassius auratus*	Copper sulfate	0.5–1.0 ppm	I	?	As needed	Effective	Use with caution in soft waters	Osborn, 1966

,,	,,	Trout	Formalin	400 ppm	F 15 min.	1	....	....	Killed fish	Leger, 1909
,,	,,	*Tinca tinca*	Formalin	Less than 400 ppm	F 15 min.	1	....	Effective	....	Roth, 1910
,,	,,	....	Formalin	....	....	....	....	....	....	Zschiesche, 1910
,,	,,	Trout	Formalin	350 ppm	F 15 min.	1	....	Effective	....	Plehn, 1924
,,	,,	Trout	Formalin	166 ppm	F 1 hr.	....	As needed	Effective	Alternate weeks	Fish, 1940
,,	,,	Unidentified	Formalin	200–500 ppm	D 30–45 min.	1	....	Effective	....	Schäperclaus, 1954
,,	,,	Unidentified	Formalin	1,000 ppm	D 15 min.	1	....	Effective	....	Schäperclaus, 1954
,,	,,	*Oncorhynchus* sp.	Furanace	1 ppm	I	?	?	Not effective	....	Amend, 1969
,,	,,	*Ictalurus punctatus*	Gentian violet	0.3 ppm	I	1	....	Inhibitory	Toxic to many species of fish	F. Meyer, Unpubl.
,,	,,	Unidentified	Globucid	2,000 ppm	F 24 hrs.	1	....	Effective	....	Schäperclaus, 1954
,,	,,	*Cyprinus carpio*	Lysol	200 ppm	D 30 sec.	1	....	Effective	Not as good as formalin	Schäperclaus, 1954
,,	,,	Trout, *Cyprinus carpio*	Malachite green oxalate	0.1–0.15 ppm	I	2–3	Alternate days	Effective	....	Amlacher, 1961b
,,	,,	Unidentified	Methylene blue	3 ppm	I	1	....	Effective	....	Amlacher, 1961b
,,	,,	*Carassius vulgaris*	Micropur	10 ppm	F 24 hrs.	1	....	Effective	....	Schäperclaus, 1954
,,	,,	Salmonids	PMA	2 ppm	F 1 hr.	1	....	Effective	Toxic to *Salmo gairdneri*	Burrows and Palmer, 1949
,,	,,	*Ictalurus punctatus*	PMA	2 ppm	F 1 hr.	1	....	Effective	....	Clemens and Sneed, 1959
,,	,,	Unidentified	Potassium permanganate	10 ppm	F 1 hr.	?	?	Effective	Toxic to *Stizostedion* sp.	Fish, 1933
,,	,,	Unidentified	Potassium permanganate	10 ppm	F 90 min.	?	As needed	Effective	....	Amlacher, 1961b

[*continued*

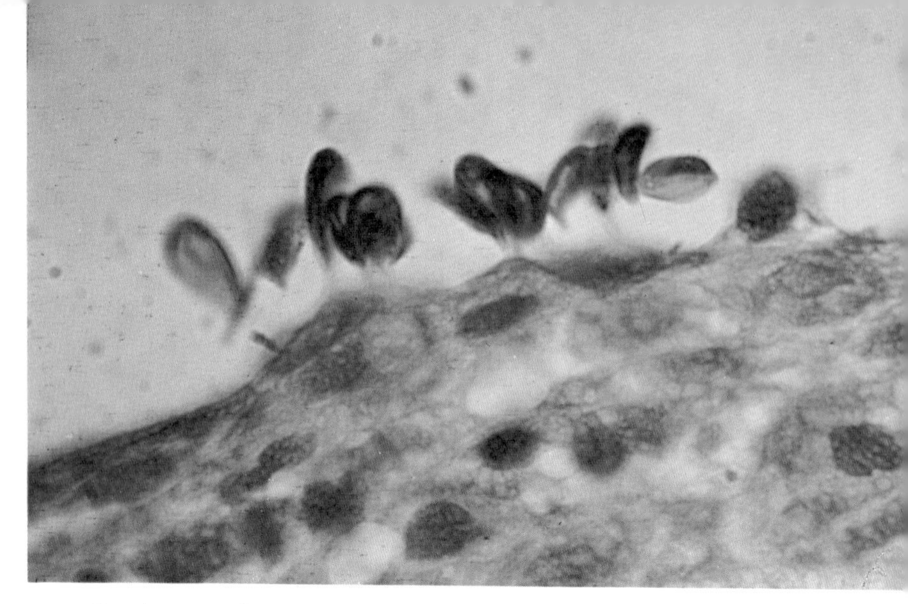

Costia necatrix on skin of brown trout. Histological section.

Cryptobia salmositica, blood flagellate from rainbow trout. Note anterior and posterior flagella. Courtesy of Dr. C. D. Becker, Battelle-Northwest, Richland, Washington.

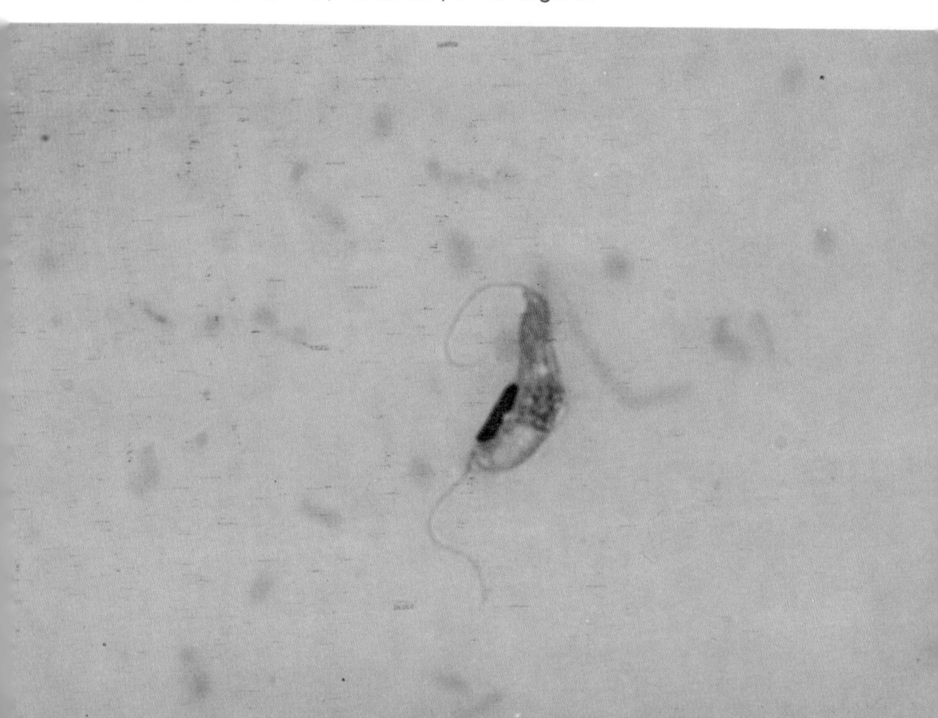

Cryptocaryon irritans in skin of *Pomacanthus semicirculatus.* Note white patches. Photo by Frickhinger.

Microscopic preparation of *Cryptocaryon irritans,* a widespread ciliate parasite of marine fishes. Note elongate nucleus. Photo by Frickhinger.

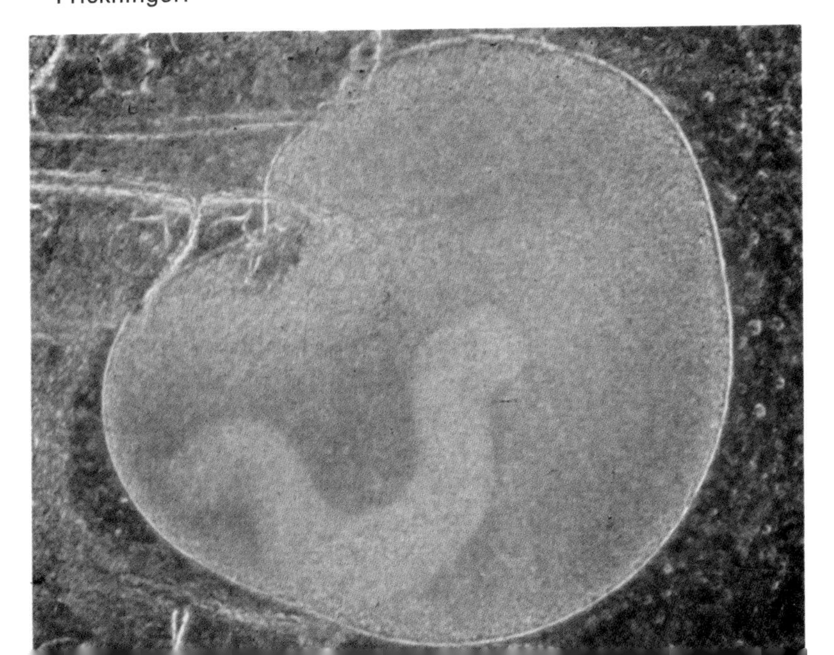

Table 3. Protozoans (External)—*continued*

Parasite	Host	Treatment	Dosage	Method	Number of Appli-cations	Frequency	Author's Report of Success	Remarks	References
Costia sp.	*Cyprinus carpio*	Potassium permanganate	1,000 ppm	D 30–45 sec.	?	As needed	Effective	....	Schäperclaus, 1954; Amlacher, 1961b; Reichenbach-Klinke, 1966
" "	Pond fishes	Potassium permanganate	1,000 ppm	F 10 min.	As needed	Daily	Effective	....	Amlacher, 1961b
" "	Unidentified	Potassium permanganate	10 ppm	F 30 min.	?	As needed	Effective	....	Reichenbach-Klinke, 1966
" "	*Cyprinus carpio*	Quicklime	2,000 ppm	D 5 sec.	1	....	Effective	....	Schäperclaus, 1954
" "	*Cyprinus carpio*	Quinine hydrochloride	20 ppm	F 24 hrs.	1	....	Effective	Toxic to fish	Schäperclaus, 1954
" "	*Cyprinus carpio*	Sodium chloride	10,000 ppm	D 15–30 min.	As needed	Daily	Effective	May recur	Schäperclaus, 1954
" "	*Cyprinus carpio*	Sodium chloride	175,500 ppm	D 3 min.	As needed	Daily	Effective	May recur	Schäperclaus, 1954
" "	*Cyprinus carpio*	Sodium chloride	25,000 ppm	D 30 sec.	As needed	Daily	Effective	May recur	Schäperclaus, 1954
" "	Unidentitied large fish	Sodium chloride	25,000 ppm	F 10–15 mins.	As needed	Daily	Effective	....	Amlacher, 1961
" "	*Cyprinus carpio*	Temperature	Raise to 32°C	P 5 days	1	....	Effective	Parasite cannot live above 30°C	Schäperclaus, 1954
" "	*In vitro*	Ultraviolet light	212,400 MWS/cm^2	P	1	....	Effective	....	Vlesanko, 1969

Cryptobia sp. (in blood)	*Cyprinus carpio*	Methylene blue	100 ppm	F 20 hrs.	1	....	Effective	Also may be fed	Havelka *et. al.*, 1965
Cryptocaryon irritans	Marine aquarium fishes	Acriflavine	10 ppm	I	1 or 2	?	Effective	....	DeGraaf, 1962
,, ,,	Marine aquarium fishes	Copper sulfate + citric acid + methylene blue	0.15 ppm $CuSO_4$ + ? citric acid + 0.2 ppm methylene blue	I	3	5 day interval	Effective	Amount of citric acid not given	Nigrelli and Ruggieri, 1966
,, ,,	Marine aquarium fishes	Cupric acetate + formalin + Tris Buffer	0.42 ppm Curpic acetate + 5.26 ppm formalin + 4.8 ppm Tris buffer	I	1	....	Usually effective	....	Nigrelli and Ruggieri, 1966
,, ,,	Marine aquarium fishes	Quinine hydro-chloride	13–20 ppm	I	As needed	4 day intervals	Effective	Change water after second treatment	DeGraaf, 1962
Cyclochaeta sp. (See also *Trichodina*)	Pond fishes	Acetic acid, glacial	2,000 ppm	D	?	....	Effective	....	Hora and Pillay, 1962
Cyclochaeta sp.	Unidentified	Albucid (sodium)	2,000 ppm	F 24 hrs.	?	....	Inhibitory	....	Schäperclaus, 1954
,, ,,	*Carassius vulgaris*	Collargol	10 ppm	F 24 hrs.	1	....	Not effective	Toxic to fish	Schäperclaus, 1954
,, ,,	Pond fishes	Formalin	400 ppm	F 1 hr.	?	....	Effective	Would kill many fishes	Hora and Pillay, 1962
,, ,,	Unidentified	Globucid (sodium)	2,000 ppm	F 24 hrs.	?	....	Inhibitory	Required 72 hrs.	Schäperclaus, 1954
,, ,,	*Carassius vulgaris*	Micropur	10 ppm	F 24 hrs.	1	....	Effective	....	Schäperclaus, 1954
,, ,,	Unidentified	Quinine hydro-chloride	20 ppm	F 6 hrs.	?	?	Effective	Toxic to fish	Schäperclaus, 1954
,, ,,	Pond fishes	Sodium chloride	30,000 ppm	F 5–10 min.	1	....	Effective	....	Horay and Pillay, 1962
Epistylis sp.	*Procambarus clarkii* (crayfish)	Formalin	50 ppm	I	1	....	Not effective	Not toxic to crayfish	F. Meyer, Unpubl.

[*continued*

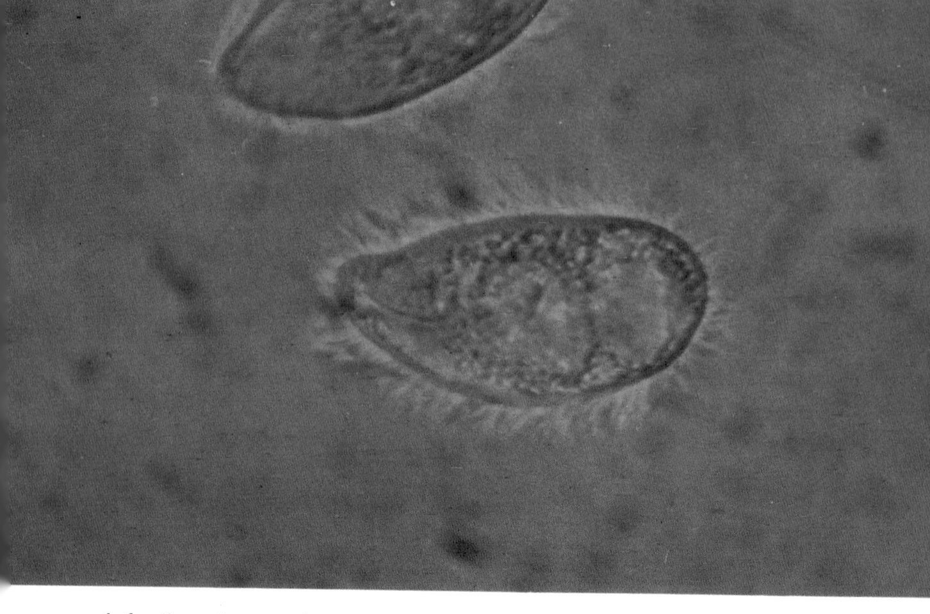

Infective, free-swimming stage of *Cryptocaryon irritans.* Courtesy of Drs. R. F. Nigrelli and G. D. Ruggieri, Osborne Laboratories of Marine Sciences, New York Aquarium.

As above, but different individual. Courtesy of Drs. R. F. Nigrelli and G. D. Ruggieri, Osborne Laboratories of Marine Sciences, New York Aquarium.

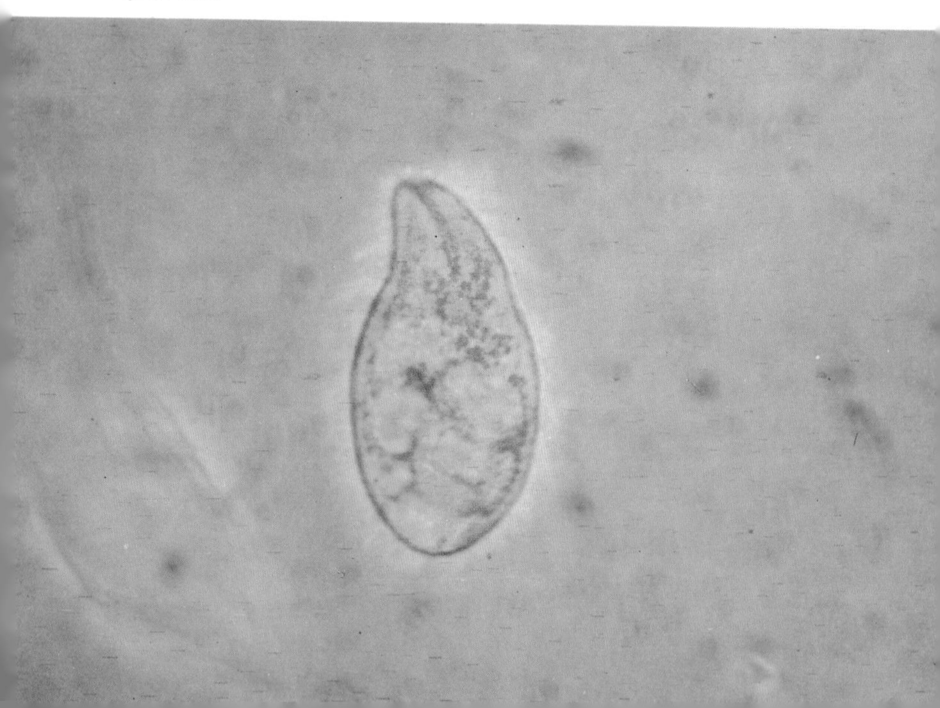

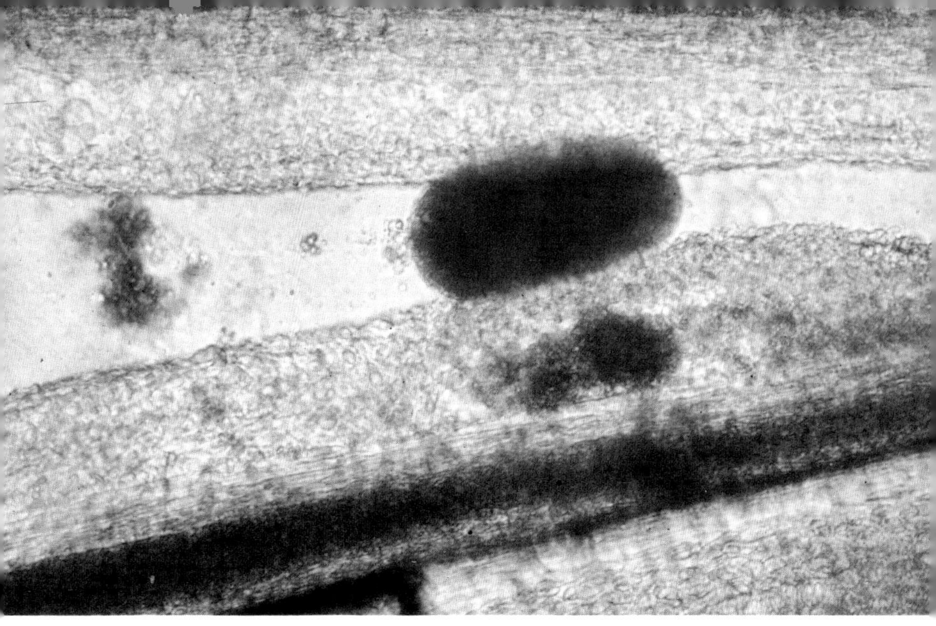

Cryptocaryon irritans trophont on the gills of aquarium fish. Courtesy of Drs. R. F. Nigrelli and G. D. Ruggieri, Osborne Laboratories of Marine Sciences, New York Aquarium.

Epistylis sp. colonies growing on warmouth bass. Courtesy of Dr. W. Rogers, Auburn University.

Table 3. Protozoans (External)—*continued*

Parasite	Host	Treatment	Dosage	Method	Number of Appli-cations	Frequency	Author's Report of Success	Remarks	References
Epistylis sp.	*Lepomis macrochirus*	Formalin	25 ppm	I	1	....	Not effective	....	F. Meyer, Unpubl.
,, ,,	Salmonids	PMA	2 ppm	F 1 hr.	1	....	Effective	Toxic to Salmo gairdneri	Burrows and Palmer, 1949
,, ,,	*Lepomis macrochirus*	Potassium permanganate	2 ppm	I	1	....	Not effective	....	F. Meyer, Unpubl.
,, ,,	Trout	Sodium chloride + malachite green	30,000 ppm NaCl + 66 ppm malachite green	D	1	....	Effective	....	Fischthal, 1949
Ichthyophthirius multifiliis	*Cyprinus carpio*	Acriflavine, neutral	10 ppm	I 3 to 20 days	I	....	Effective	....	Schäperclaus, 1954 Bauer, 1958
,, ,,	*Ictalurus punctatus*	Acriflavine, neutral	3 ppm	I	1	....	Not effective	....	F. Meyer, Unpubl.
,, ,,	*In vitro*	Acriflavine, neutral	3 ppm	I	1	....	Effective	1 and 2 ppm not effective; fish not included	Hoffman and Putz, 1966
,, ,,	Pond fishes	Acriflavine, neutral	1 ppm	I	1	....	Not effective	Used to control bacteria in shipping	Swingle, 1955
,, ,,	Unidentified	Aluminum sulfate	50,000 ppm	D 1 min.	1	....	Effective	....	Gopalakrishnan, 1966 Prytherch, 1928
,, ,,	*In vitro*	Amopyroquin	0.5 ppm	I	1	....	Not effective	....	Hoffman and Putz, 1966
,, ,,	*In vitro*	Amprolium	1 ppm	I	1	....	Not effective	....	Hoffman and Putz, 1966

,,	,,	*In vitro*	Aqua-aid	2× recommended on label	I	1	....	Effective	Recommended dosage not effective	Hoffman and Putz, 1966
,,	,,	Unidentified	Aquarol	80 ppm	I	3	3 days	Effective	....	Amlacher, 1961b; Reichenbach-Klinke, 1966
,,	,,	Aquarium fishes	Atebrine	1.6 ppm	I	3	Alternate days	Effective	Toxic to *Lebistes* sp.	Slater, 1952
,,	,,	Unidentified	Atebrine	2.5 ppm	F	?	?	Effective	Toxic to *Lebistes* sp.	Schäperclaus, 1954
,,	,,	Aquarium fishes	Atebrine	?	....	....	....	Effective	....	Pfeiffer, 1952
,,	,,	Unidentified	Atebrine	?	....	....	....	Effective	....	Chang, 1960
,,	,,	Unidentified	Atebrine	10 ppm	I	?	?	Effective	....	Amlacher, 1961b; Reichenbach-Klinke, 1966
,,	,,	Aquarium fishes	Atebrine	3.5 ppm	I	3	Alternate days	Effective	....	Van Duijn, 1967
,,	,,	Unidentified	Aureomycin	13 ppm	I	?	?	Effective	....	Amlacher, 1961b
,,	,,	*Salmo gairdneri*	Aureomycin	0.001–0.0015 ppm in feed	0	7	1–4 days	Effective	....	Borshosh and Illesh, 1962
,,	,,	Unidentified	Aureomycin	13 ppm	I	1	....	Effective	Effect may be on a secondary bacterial infection	Reichenbach-Klinke, 1966
,,	,,	*In vitro*	Basic bright green (brilliant green)	0.12–1 ppm	I 2.5–5 hrs.	1	....	Effective	....	Musselius and Filippova, 1968, 1969
,,	,,	Trout	Basic bright green (brilliant green)	0.25–0.5 ppm	F 5 hrs.	1	....	Trophs killed sub-epithelially	18 × cheaper than malachite green	Musselius and Filippova, 1968, 1969

[*continued*

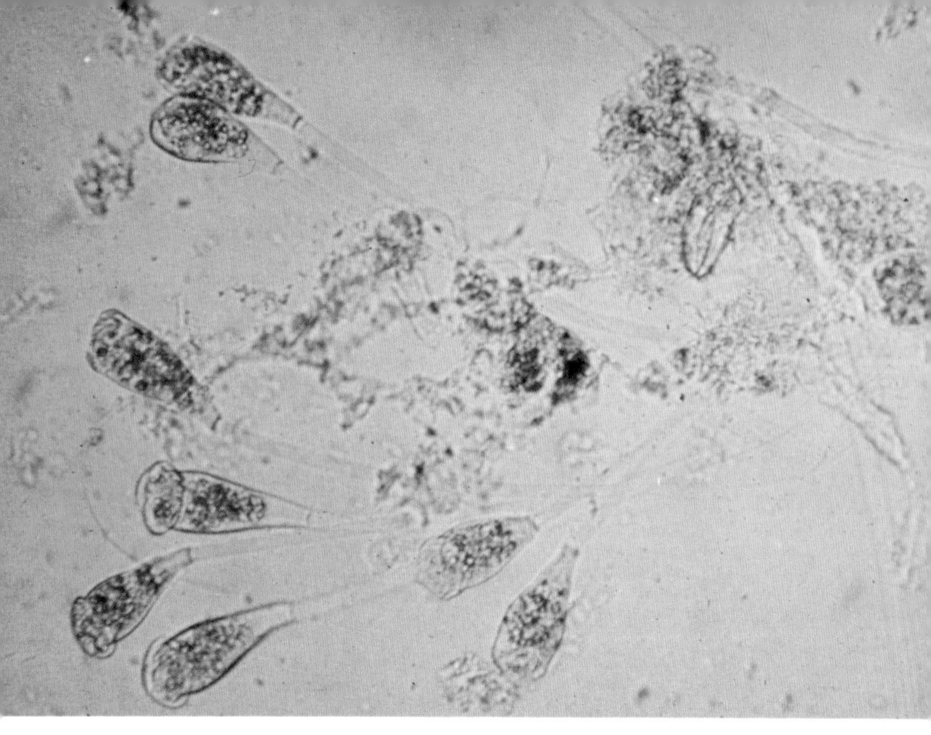

Epistylis sp., living colony from largemouth bass. Courtesy of Dr. W. Rogers, Auburn University.

Ichthyophthirius multifiliis (white spot) in skin of a 3-spined stickleback.

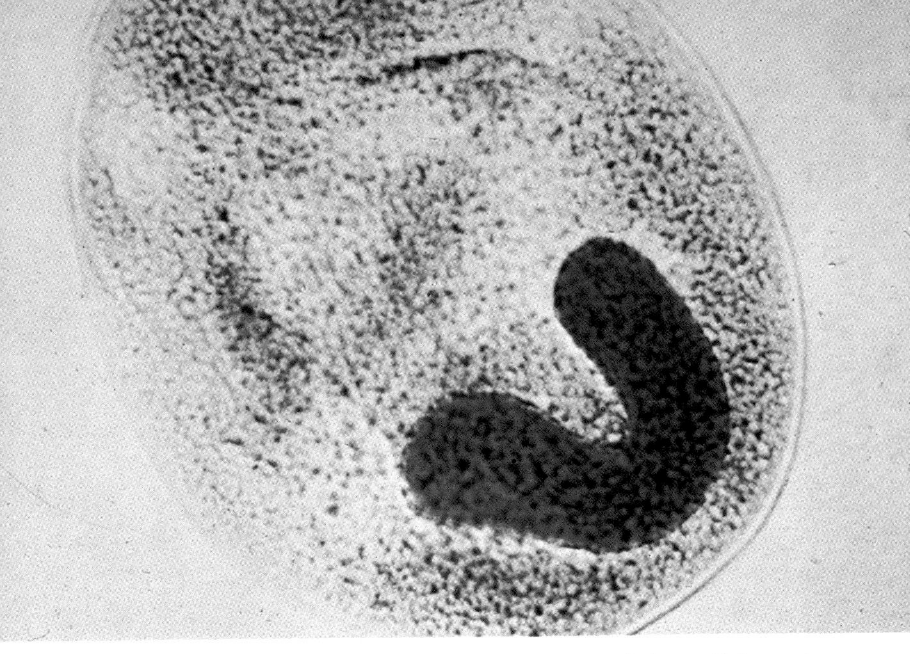

Ichthyophthirius multifiliis, mature form removed from fish and stained. Note horseshoe-shaped nucleus. Photo by F. Meyer.

Ichthyophthirius multifiliis, tomite (small invasive stage). Note tear-drop shape with pointed anterior end. Courtesy of R. Whitman, Beckley, WV.

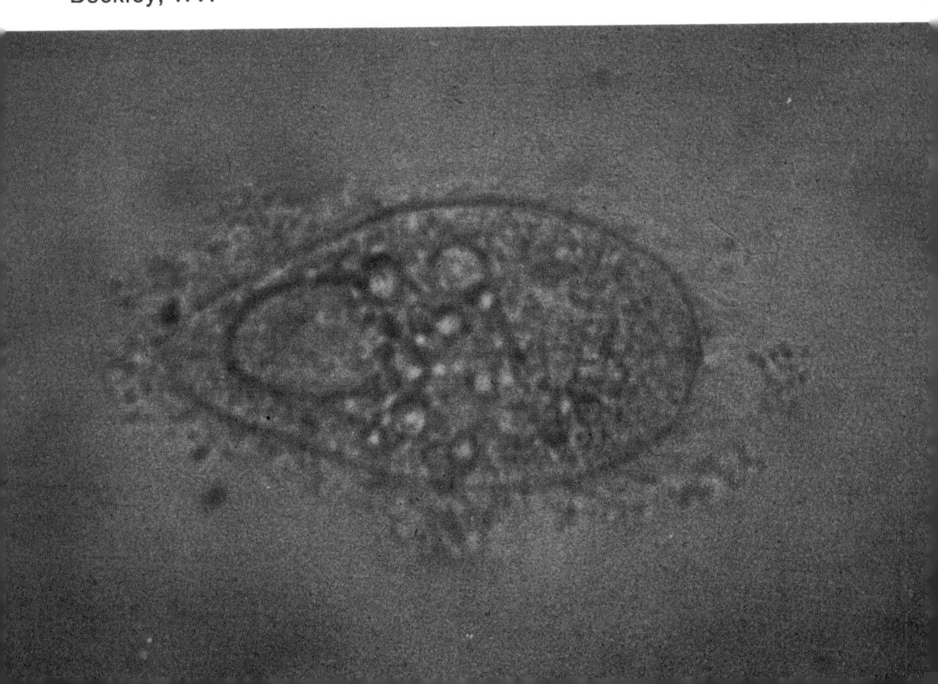

Table 3. Protozoans (External)—*continued*

Parasites	Host	Treatment	Dosage	Method	Number of Appli-cations	Frequency	Author's Report of Success	Remarks	References
Ichthyophthirius multifiliis	*In vitro*	Basic violet	0.12–0.25 ppm	I ?	1	....	Effective	....	Musselius and Filippova, 1968, 1969
,, ,,	Unidentified	Basic violet	0.12–0.25 ppm	F 5.5 hrs.	1	....	Trophs killed sub-epithelially	....	Musselius and Filippova, 1968, 1969
,, ,,	Trout	Basic violet	0.25–0.5 ppm	F 2–4 hrs.	1	....	Trophs killed sub-epithelially	....	Musselius and Filippova, 1968, 1969
,, ,,	*In vitro*	Betadine	10 ppm	I	1	....	Not effective	Fish not included	Hoffman and Putz, 1966
,, ,,	*Cyprinus carpio*	Basic bright green	0.25–0.5 ppm	F 5 hrs.	1	....	Effective	....	Musselius and Filippova, 1968
,, ,,	Aquarium fishes	Chloramine-B (Halamid)	10 ppm	F 24 hrs.	1	....	Effective	Do not use in metal tanks	Van Duijn, 1967
,, ,,	*Ctenopharyngodon idellus*	Chloramine-T	20 ppm	I	14	Twice daily	Effective	Soft or hard waters, pH 7	Cross, 1971
,, ,,	*Ctenopharyngodon idellus*	Chloramine-T	5 ppm	I	14	Twice daily	Effective	Soft water, pH 6	Cross, 1971
,, ,,	*Ictalurus punctatus*	Chelated copper ($CuSo_4$ + EDTA)	1 ppm	I	?	Daily	Not effective	....	F. Meyer, Unpubl.
,, ,,	*Ictalurus punctatus*	Chloramphenicol	12.5 ppm	I	Several	Daily	Not effective	....	F. Meyer, Unpubl.
,, ,,	Aquarium fishes	Chloramphenicol	6–10 ppm	I	1	....	....	Reported as effective but no experiments cited	Van Duijn, 1967

,,	,,	*Salmo gairdneri*	Chlortetracycline (Biomycin)	1,000 ppm in feed	O	7	1–4 days	Effective	Effect may have been on secondary bacteria	Borshosh and Illesh, 1962
,,	,,	Trout	Copper sulfate The biological implications of using copper sulfate for algal control were reviewed by Moyle (1949) and Nichols *et al.* (1946)	500 ppm plus 500 ppm of acetic acid	D 1–2 min.	As needed	Daily	Effective	Use with extreme caution if carbonate in the water is less than 50 ppm	Leitritz, 1960
,,	,,	*Ictalurus punctatus*	Copper sulfate	0.25–2 ppm add 3 ppm acetic acid in hard waters	I	As needed	Alternate days or weekly	Good for prophylaxis	Use with extreme caution, if 50 to 200 ppm carbonate, do not exceed 1 ppm; if over 200 ppm can use more	Bishop, 1963; Maloy, 1966
,,	,,	*Ictalurus punctatus*	Copper sulfate	1 ppm	I	Several	Daily	Not effective	....	F. Meyer, Unpubl.
,,	,,	*Ictalurus punctatus*	Copper sulfate	2 ppm	I	Several	Daily	?	Toxic to fish	F. Meyer, Unpubl.
,,	,,	*Ictalurus punctatus*	Copper sulfate + Citric acid	1 ppm 2 ppm	I	Several	Alternate days	Not effective	....	F. Meyer, Unpubl.
,,	,,	*Ictalurus punctatus*	Copper sulfate + DMSO	0.1–1.0 ppm + 0.5–5 ppm	I	....	Alternate days	Not effective	....	F. Meyer, Unpubl.
,,	,,	*Ictalurus punctatus*	Co-Ral	1 ppm	I	1	....	Not effective	....	R. Allison, 1969
,,	,,	*In vitro*	Daraprim	1 ppm	I	1	....	Not effective	....	Hoffman and Putz, 1966
,,	,,	*Ictalurus punctatus*	Dylox	1–2 ppm	I	Several	Daily	Not effective	....	F. Meyer, 1970; F. Meyer, Unpubl.
,,	,,	*Ictalurus punctatus*	Dylox + Copper sulfate	1 ppm + 0.5 to 1 ppm	I	?	Daily	Not effective	....	F. Meyer, Unpubl.

[*continued*

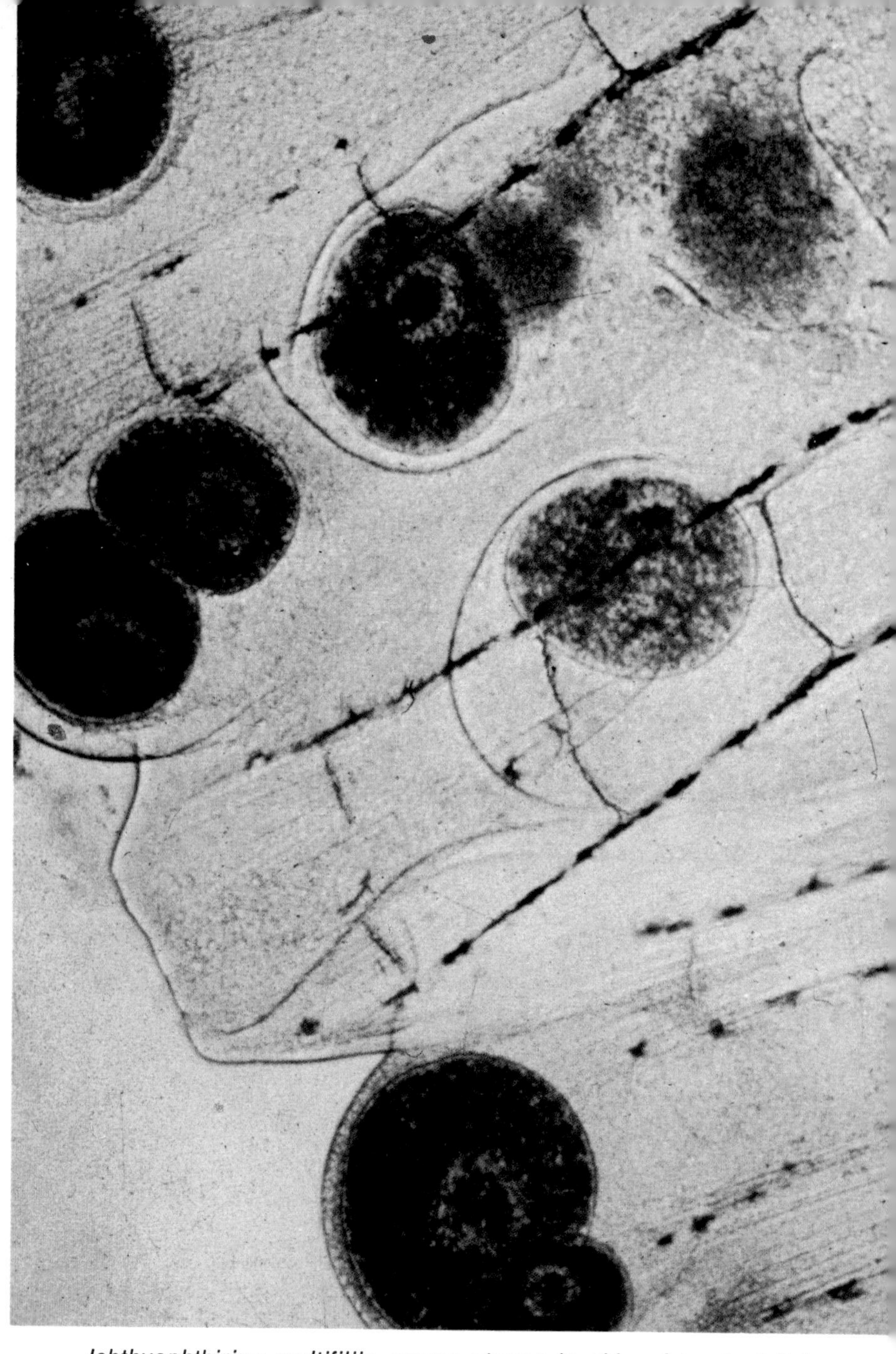

Ichthyophthirius multifiliis, young stages in skin of tropical fish. Photo by Frickhinger.

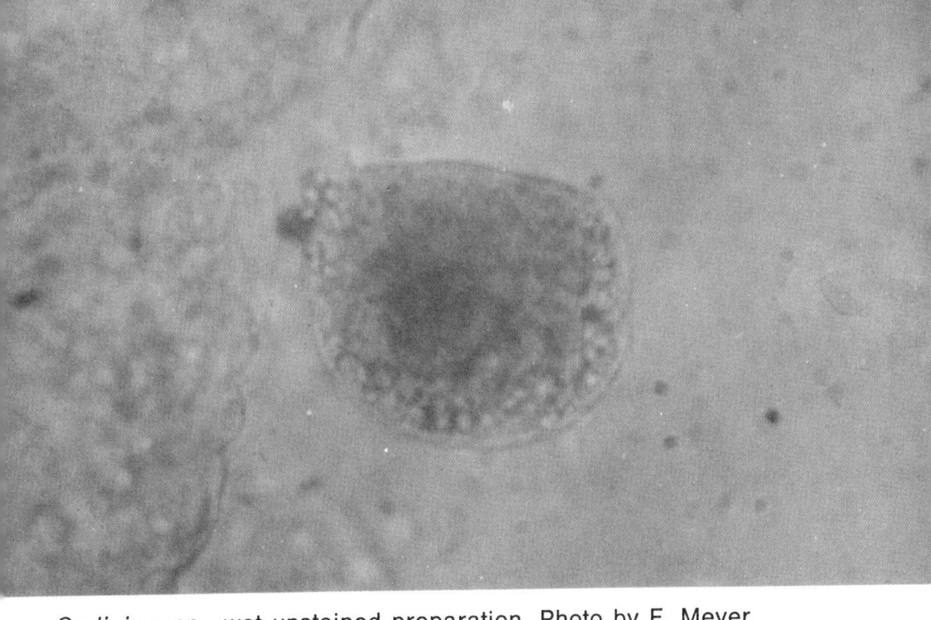

Oodinium sp., wet unstained preparation. Photo by F. Meyer.

Tetrahymena corlissi facultative parasite from guppy epizootic. Note pear shape. From Hoffman et al., manuscript in preparation. Photo by J. Camper.

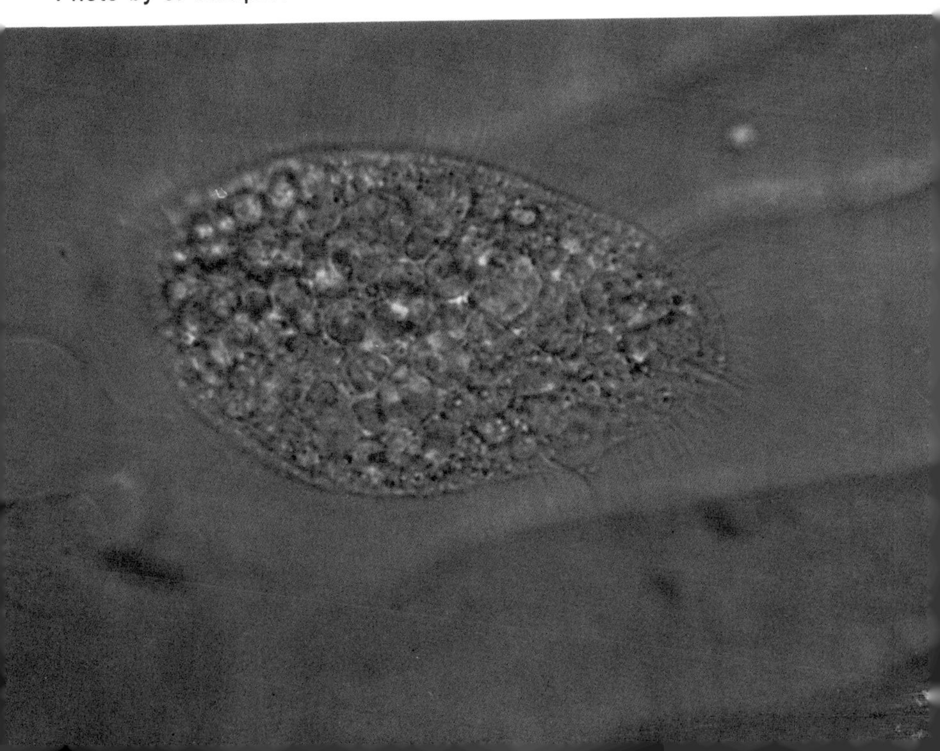

Table 3. Protozoans (External)—*continued*

Parasite	Host	Treatment	Dosage	Method	Number of Appli-cations	Frequency	Author's Report of Success	Remarks	References
Ichthyophthirius multifiliis	*Ictalurus punctatus*	Dylox + Copper sulfate	1 ppm 2 ppm	I	?	Daily	....	Toxic to fish	F. Meyer, Unpubl.
,, ,,	*Ictalurus punctatus*	DMSO	2,500 ppm	I	1	....	Not effective	....	F. Meyer, Unpubl.
,, ,,	*Ictalurus punctatus*	Ethylene glycol	1,000 ppm	I	5	Daily	Not effective	....	F. Meyer, Unpubl.
,, ,,	*In vitro*	Enheptin	10 ppm	I	1	....	Not effective	....	Hoffman and Putz, 1966
,, ,,	Unidentified	Eosin	60 ppm	I 5 days	....	....	Effective	Killed free parasites in 16 to 60 minutes	Stiles, 1928
,, ,,	*Salmo gairdneri*	Flagyl	10,000 ppm in feed	O	28	Daily	Effective	28 days probably too long; parasites reduced in 5 days	Hoffman and Putz, 1966
,, ,,	*In vitro*	Flagyl	1.5 ppm	I	1	....	Inhibitory	Killed trophs and tomites	Hoffman and Putz, 1966
,, ,,	*Ictalurus punctatus*	Flagyl	10,000 ppm in feed	O	?	Daily	Unknown	Sick fish refused food	F. Meyer, Unpubl.
,, ,,	*Ictalurus punctatus*	Flagyl	25 ppm	I	?	Daily	Not effective	....	F. Meyer, Unpubl.
,, ,,	Trout	Formalin	333–500 ppm	Constant flow, time not given	Not given	Not given	Effective	Eradicated monogenea and protozoan parasites	Kingsbury and Embody, 1932
,, ,,	Pond fishes	Formalin	200 ppm	F 1 hr.	7–10	Daily	Effective	....	Gopalakrishnan, 1963 and 1964
,, ,,	Pond fishes	Formalin	400 ppm	F 15 min.	?	Daily	Effective	....	Gopalakrishnan, 1963 and 1964
,, ,,	Salmonids	Formalin	200–250 ppm	F 1 hr.	As needed	Daily	Effective	Use less if water is above 60°F	Davis, 1953

,,	,,	Pond fishes	Formalin	166–200 ppm	F 1 hr.	As needed	Daily	Effective	Use less when water is warmer	Wellborn, 1965; F. Meyer, 1966a
,,	,,	Pond fishes	Formalin	15–25 ppm	I	1	....	Effective	Avoid use during extreme hot weather	R. Allison, 1957a, 1962, 1963
,,	,,	*Ictalurus punctatus*	Formalin	15 ppm	I	4	Alternate days	Effective	Causes oxygen depletion during hot weather	Wellborn, 1965
,,	,,	*Ictalurus punctatus*	Formalin	15–25 ppm	Constant flow	1	....	Effective	Use until danger is past	Mount, 1965
,,	,,	*In vitro*	Formalin	10–25 ppm	I	1	....	Effective	Killed trophs and tomites; fish not included	Hoffman and Putz, 1966
,,	,,	Trout	Formalin	30 ppm	F 6 hrs.	14	Daily	Effective	No ''Ich'' could be found 2 months later	Camper, 1970
,,	,,	*Ictalurus punctatus*	Formalin	200 ppm	F 1 hr.	14	Daily	Effective	....	Meyer and Collar, 1964
,,	,,	*Ictalurus punctatus*	Formalin	15 ppm	I	3	Alternate days	Not effective	....	F. Meyer, Unpubl.
,,	,,	*Ictalurus punctatus*	Formalin	25 ppm	I	3	Alternate days	Effective	If treated early	F. Meyer, Unpubl.
,,	,,	*Ictalurus punctatus*	Formalin	50 ppm	I	3	Alternate days	Effective	Toxic to weak fish	F. Meyer, Unpubl.
,,	,,	*Ictalurus punctatus*	Formalin + DMSO	10 ppm 50 ppm	I	?	Alternate days	Not effective	....	F. Meyer, Unpubl.
,,	,,	*Ictalurus punctatus*	Formalin + DMSO	25 ppm 125 ppm	I	?	Alternate days	Inhibitory	....	F. Meyer, Unpubl.
,,	,,	*Ictalurus punctatus*	Formalin + DMSO	50 ppm 250 ppm	I	?	Alternate days	Effective	Slow to control	F. Meyer, Unpubl.

[*continued*

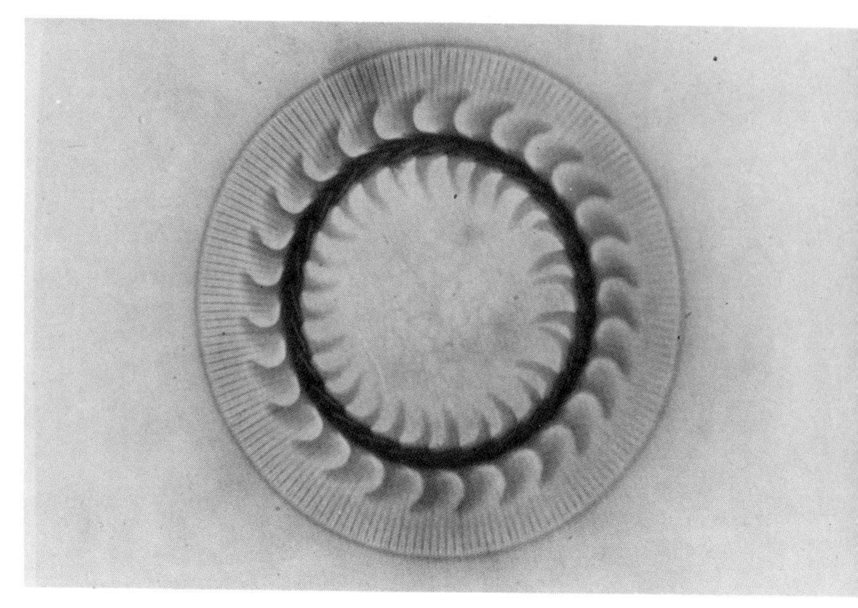

Trichodina fultoni. Note striated ring and the inner ring of denticles composed of central thorns, center pieces, and outer blades. Photo by the late Dr. H. S. Davis.

Trichodina sp. on gills of channel catfish. Photo by F. Meyer.

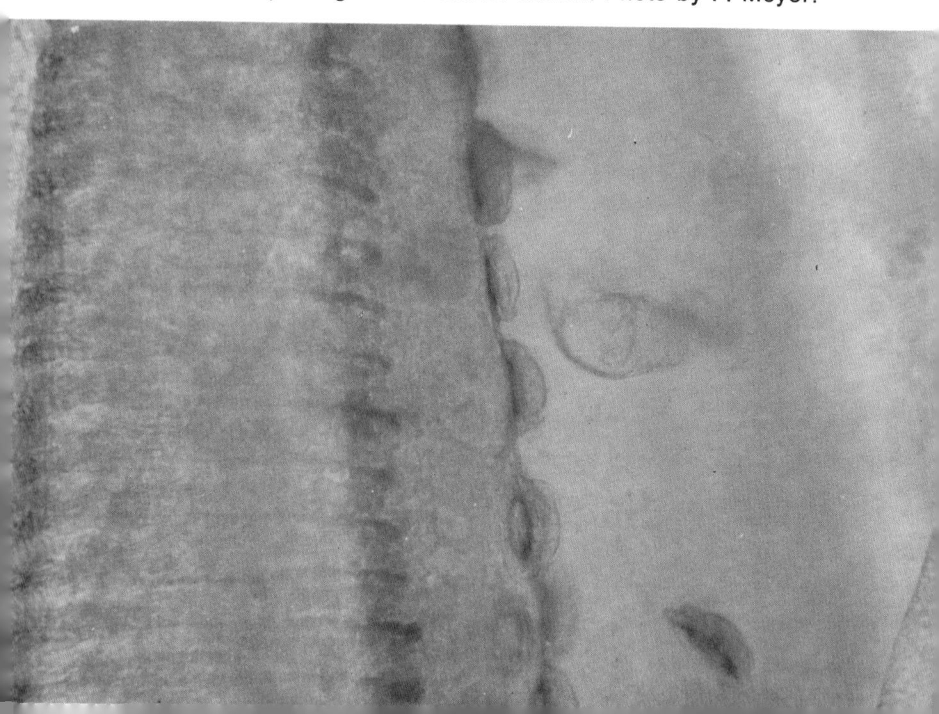

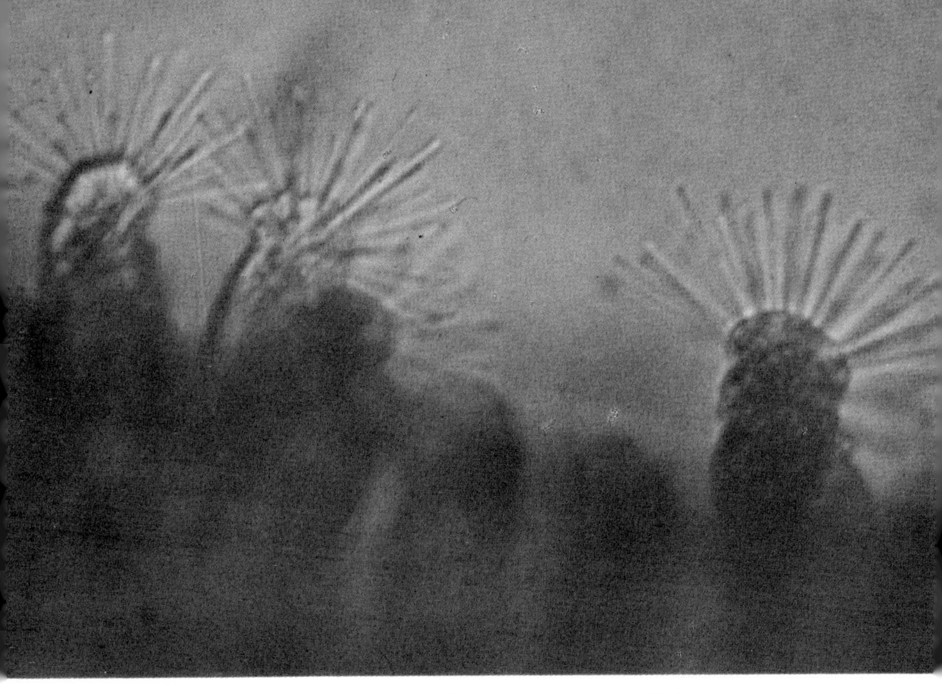

Trichophrya piscium on gills of Atlantic salmon. Note extended food-gathering tentacles. Courtesy of R. Dexter, USF & WS.

Ceratomyxa shasta causing fatal boil-like disease of coho salmon. Courtesy of J. Conrad, Fish Commission of Oregon.

Table 3. Protozoans (External)—*continued*

Parasite	Host	Treatment	Dosage	Method	Number of Applications	Frequency	Author's Report of Success	Remarks	References
Ichthyophthirius multifiliis	*In vitro*	Furacin	10 ppm	I	1	....	Not effective	Trophs and tomites	Hoffman and Putz, 1966
,, ,,	Aquarium fishes	Halamid	10 ppm	I	?	Weekly	Effective	....	Postema, 1956
,, ,,	*Ictalurus punctatus*	Kanamycin	12.5 ppm	I	?	Daily	Not effective	....	F. Meyer, Unpubl.
,, ,,	Pond fishes	Lilac leaves	Not given	I	....	....	....	Anti-protozoal action reported	Bauer, 1959
,, ,,	*Ictalurus punctatus*	Malachite green oxalate	0.1 ppm	I	1	....	Effective	In ponds	R. Allison, 1963
,, ,,	Trout; *Cyprinus carpio*	Malachite green	0.15 ppm	I	2–3	Alternate days	Effective	....	Amlacher, 1961a
,, ,,	Trout	Malachite green oxalate	1.25 ppm	F 30 min.	?	Daily	Effective	Use until "Ich" disappears	A. Johnson, 1961
,, ,,	*Cyprinus carpio*	Malachite green oxalate	0.5 ppm	I	3–6	Daily	Effective	Also used in carp feeding areas	Avdosev, 1962
,, ,,	*Cyprinus carpio*	Malachite green oxalate	0.25 ppm	I	Several	2–3 day intervals	Effective	....	Kubu, 1962
,, ,,	Pond fishes	Malachite green oxalate	0.15 ppm	I	....	....	Effective	Good review, includes toxicity	Steffens, 1962
,, ,,	Pond fishes	Malachite green oxalate	0.1 ppm	I	2	Weekly	Effective	Some claim 0.1 ppm not effective	Beckert and Allison, 1964; R. Allison, 1966; Havelka and Petrovicky, 1967
,, ,,	*Cyprinus carpio*	Malachite green oxalate	0.5–0.9 ppm	F 4–5 hrs.	2–3	Every 2nd day	Effective	....	Ivasik and Svirepo, 1964

,,	,,	*Cyprinus carpio*	Malachite green oxalate	4–7.5 ppm	F 1–1½ hrs.	2	Twice	Effective	....	Kocytowski and Antychowicz, 1964
,,	,,	*In vitro*	Malachite green oxalate	0.1 ppm	I	....	....	Effective	Consistently killed trophs and tomites	Hoffman and Putz, 1966
,,	,,	Unidentified	Malachite green oxalate	2 ppm	F 30 min.	As needed	Daily	Effective	No parasites seen after 5 days	Peterson, Steucke and Lynch, 1966
,,	,,	*Siluris glanis* fry	Malachite green	0.15 ppm	I	....	As needed	Effective	....	Tesarcik and Mares. 1966
,,	,,	*Cyprinus carpio*	Malachite green	4–7.5 ppm	F 1–1½ hrs.	....	Repeat	Effective	....	Tesarcik and Havelka, 1966
,,	,,	*Cyprinus carpio*	Malachite green	0.5 ppm	I	....	Weekly	Effective	Effective if used early	Tesarcik and Havelka, 1966
,,	,,	*Cyprinus carpio*	Malachite green	0.15 ppm	I	....	As needed	Effective	....	Sarig, 1968
,,	,,	Salmon	Malachite green	....	....	....	....	....	Studied effect on blood	Glagoleva and Malikova, 1968
,,	,,	Pond fishes	Malachite green	1.5 ppm	I	1	6–24 hrs.	Effective	....	Amlacher, 1961a
,,	,,	*Cyprinus carpio*	Malachite green	0.9 ppm	F 5 hrs.	3	Alternate days	Effective	Was best of 9 methods tested	Prost and Studnicke, 1971
,,	,,	*Ictalurus punctatus*	Malachite green	0.05 ppm	I	3	Alternate days	Not effective	....	Leteux and Meyer, 1972
,,	,,	*Ictalurus punctatus*	Malachite green	0.1 ppm	I	3	Alternate days	Not effective	Suppressed disease; lethal to fish	Leteux and Meyer, 1972
,,	,,	*Ictalurus punctatus*	Malachite green	0.2 ppm	I	3	Alternate days	Killed parasites	Suppressed disease; lethal to fish	Leteux and Meyer, 1972
,,	,,	*Ictalurus punctatus*	Malachite green + Formalin	0.05 ppm + 15 ppm	I	3	Alternate days	Fair	Must treat early	Leteux and Meyer, 1972
,,	,,	*Ictalurus punctatus*	Malachite green + Formalin	0.05 ppm + 25 ppm	I	3	Alternate days	Effective	....	Leteux and Meyer, 1972

[*continued*

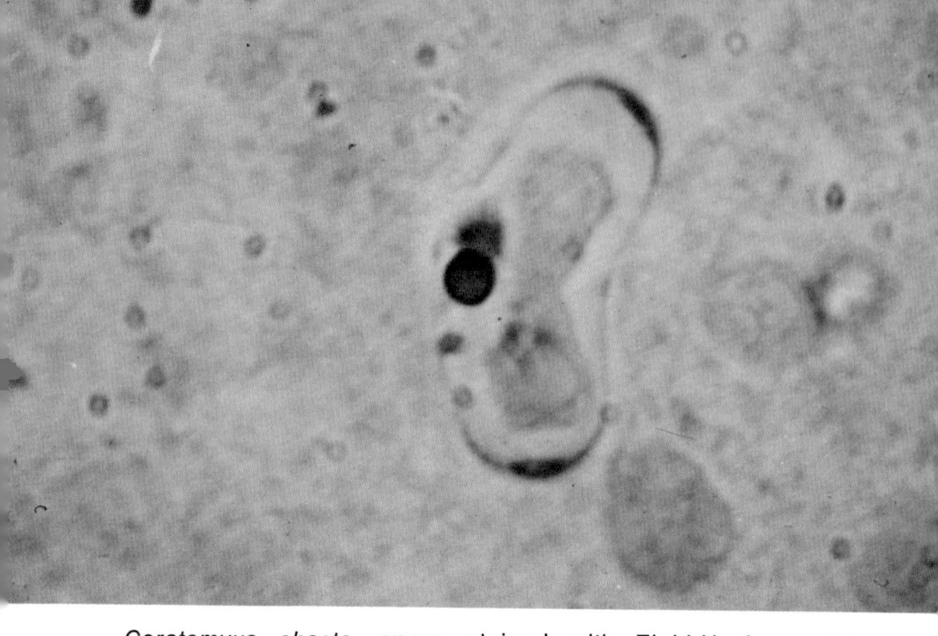

Ceratomyxa shasta, spore stained with Ziehl-Neelsen stain, showing dark blue polar capsules and light blue sporoplasm. Courtesy of J. Sanders.

Henneguya sp. causing fulminating skin lesions and hemorrhage in channel catfish. Photo by F. Meyer.

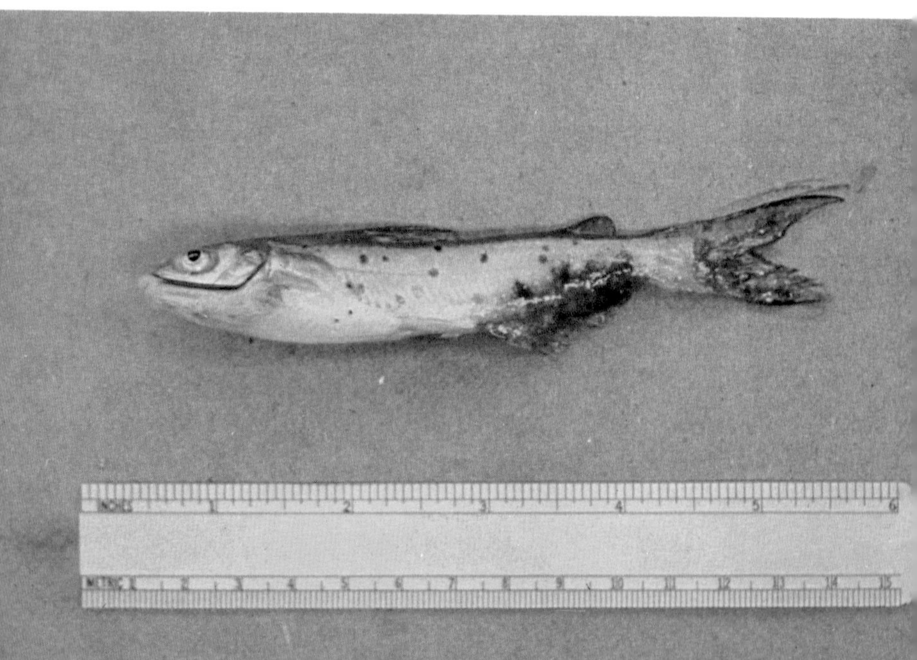

Henneguya cysts in the dorsal fin of *Leporinus*. Photo by Frickhinger.

Henneguya sp. spores, interlamellar gill form from channel catfish. Photo by F. Meyer.

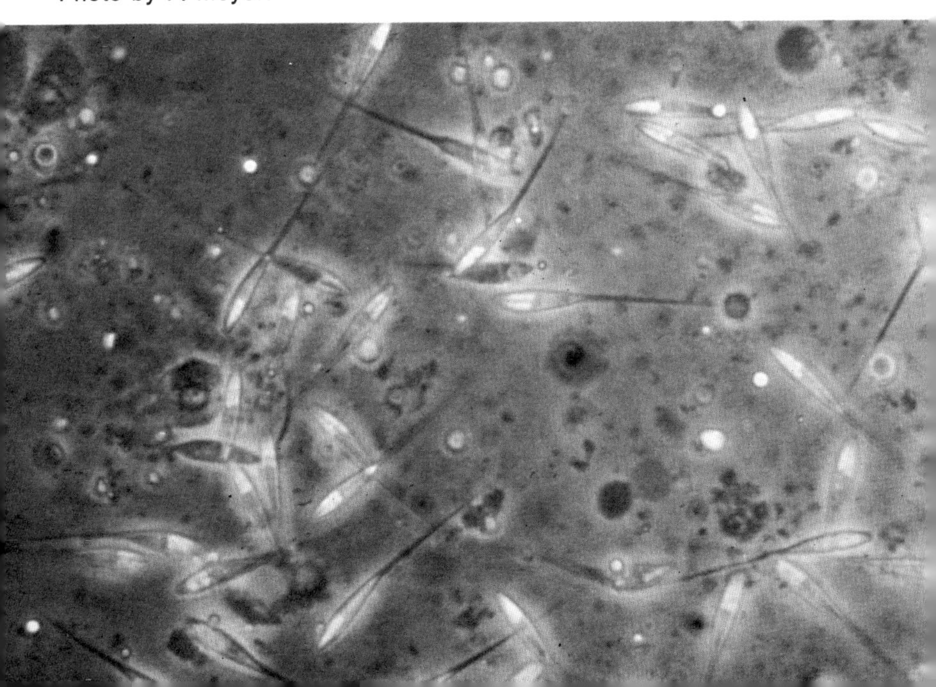

Table 3. Protozoans (External)—*continued*

Parasite	Host	Treatment	Dosage	Method	Number of appli-cations	Frequency	Author's Report of Success	Remarks	References
Ichthyophthirius multifiliis	*Ictalurus punctatus*	Malachite green + Formalin	0.05 ppm + 50 ppm	I	3	Alternate days	Effective	....	Leteux and Meyer, 1972
,, ,,	*Ictalurus punctatus*	Malachite green + Formalin	0.10 ppm + 15 ppm	I	3	Alternate days	Not effective	....	Leteux and Meyer, 1972
,, ,,	*Ictalurus punctatus*	Malachite green + Formalin	0.10 ppm + 25 ppm	I	3	Alternate days	Effective	....	Leteux and Meyer 1972
,, ,,	*Ictalurus punctatus*	Malachite green + Formalin	0.10 ppm + 50 ppm	I	3	Alternate days	Effective	Toxic to fish	Leteux and Meyer, 1972
,, ,,	*Ictalurus punctatus*	Malachite green + Formalin	0.2 ppm + 15 ppm	I	3	Alternate days	Effective	Toxic to fish	Leteux and Meyer, 1972
,, ,,	*Ictalurus punctatus*	Malachite green + Formalin	0.2 ppm + 25 ppm	I	3	Alternate days	Effective	Toxic to fish	Leteux and Meyer, 1972
,, ,,	*Ictalurus punctatus*	Malachite green + Formalin	0.2 ppm + 50 ppm	I	3	Alternate days	Effective	Toxic to fish	Leteux and Meyer, 1972
,, ,,	*Ictalurus punctatus*	Malachite green + DMSO	0.1 ppm + 0.5 ppm	I	?	Alternate days	Inhibitory	Not toxic to fish	F. Meyer, Unpubl.
,, ,,	*Ictalurus punctatus*	Malachite green + DMSO	0.5 ppm + 2.5 ppm	I	?	Alternate days	Inhibitory	Toxic to fish	F. Meyer, Unpubl.
,, ,,	*Ictalurus punctatus*	Malachite green + DMSO	1 ppm + 5 ppm	I	?	Alternate days	Unknown	Toxic to fish	F. Meyer, Unpubl.
,, ,,	*Cyprinus carpio*	Mercuric nitrate	0.3 ppm	I 4 days	1	1	Effective	At 10°C	Ivasik and Svirepo, 1964
,, ,,	*Cyprinus carpio*	Mercuric nitrate	0.2 ppm	I 4 days	1	1	Effective	At 10 to 20°C	Ivasik and Svirepo, 1964

,,	,,	*Cyprinus carpio*	Mercuric nitrate	0.1 ppm	I 4 days	1	1	Effective	At 20°C or higher	Ivasik and Svirepo, 1964
,,	,,	Aquarium fishes	Mercurochrome	Reported in drops	I	1	....	Probably effective	Toxic to fish	Thadially, 1964
,,	,,	*In vitro*	Metasol L	0.05 ppm	I	1	....	Effective	Killed trophs and tomites; fish not included	Hoffman and Putz, 1966
,,	,,	Trout, catfish	Methylene blue	Probably 3 ppm (amount not clearly stated)	I	1	....	Effective	Parasites leaving fish were killed	Stiles, 1928
,,	,,	Pond fishes	Methylene blue	3 ppm	I	1	....	Effective	....	Schäperclaus, 1954
,,	,,	Undidentified	Methylene blue	....	....	....	....	....	Reduces oxygen of water	R. Allison, 1962
,,	,,	Pond fishes	Methylene blue	1 ppm	I	1	As needed	Effective	....	R. Allison, 1966
,,	,,	*In vitro*	Methylene blue	2 ppm	I	1	....	Effective	Killed trophs and tomites	Hoffman and Putz, 1966
,,	,,	Unidentified	Methylene blue	3 ppm	I	1	....	Not effective	....	Schäperclaus, 1954
,,	,,	Aquarium fishes	Methylene blue	2–4 ppm	I	....	Daily	Effective	Higher concentration kills plants	Van Duijn, 1967
,,	,,	Aquarium fishes	Methylene blue	3 ppm	I	1	As needed	Effective	Add more to restore blue color on 3rd and 5th days	Thadially, 1963, 1964
,,	,,	Unidentified	Methylene blue	?	0	?	?	Effective	....	Ivasik *et al.*, 1967
,,	,,	*Ictalurus punctatus*	Naled (Dibrom)	1–1.5 ppm	I	As needed	Daily	Inhibitory in aquaria	Not effective in ponds	F. Meyer, Unpubl.
,,	,,	*Cyprinus carpio*	Neguvon	5,000 ppm	D 5 min	....	As needed	....	Effective but toxic to fish, lesser concentrations not effective	Prost and Studnicka, 1967

[*continued*

Hexamitiasis, rainbow trout. Note emaciation and empty intestine.

Hexamita intestinalis in an intestinal smear. Photo by Frickhinger.

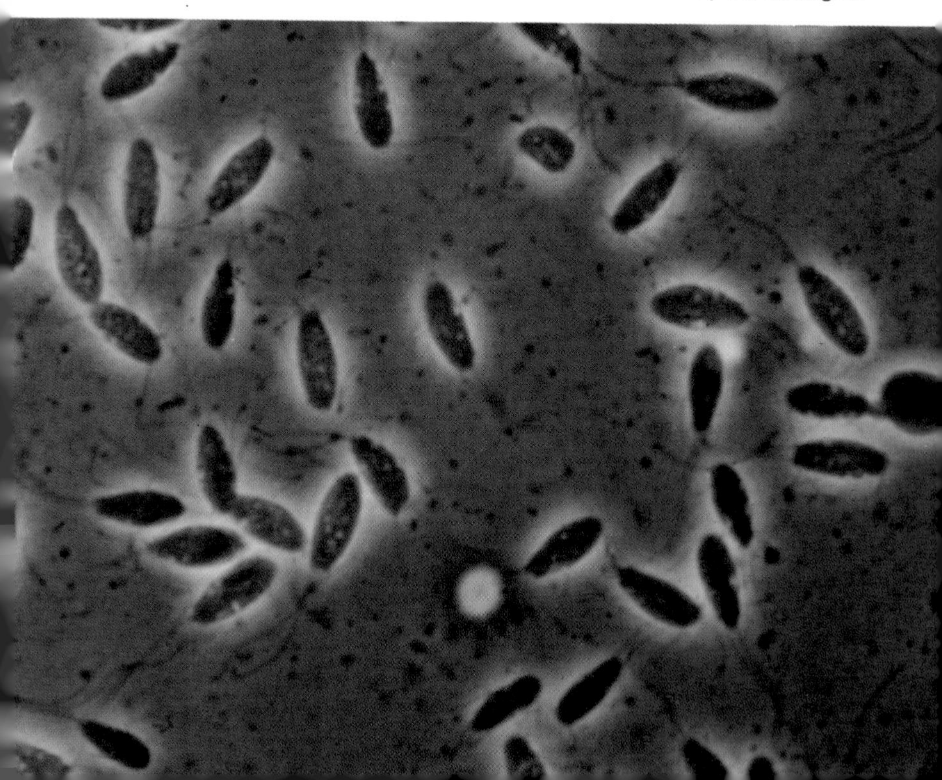

Whirling disease (*Myxosoma cerebralis*), rainbow trout. Very young stage. Note black tail. Courtesy of Dr. S. F. Snieszko, USF & WS, Eastern Fish Disease Lab., Kearneysville, W.Va.

Whirling disease, rainbow trout, three years post-infection. Experimental fish reared by W. Hill. Note extreme spinal urvature. Photo by G. Hoffman.

Table 3. Protozoans (External)—*continued*

Parasite	Host	Treatment	Dosage	Method	Number of Appli-cations	Frequency	Author's Report of Success	Remarks	References
Ichthyophthirius multifiliis	*In vitro*	Nickel sulfate	10 ppm	....	1	....	Effective	Killed trophs and tomites; fish not included	Hoffman and Putz, 1966
,, ,,	*Carassius auratus*	Nifuripirinol	1–2 ppm	D 5–10 min.	?	....	Effective	....	Takase, *et al.*, 1971
,, ,,	Aquarium fishes	Nifuripirinol	0.05–0.2 ppm	I	?	Daily	Effective	....	Takase, *et al.*, 1971
,, ,,	*Ictalurus punctaus*	Oxytetracycline	12.5 ppm	I	?	Daily	Not effective	....	F. Meyer, Unpubl.
,, ,,	Unidentified	Plasmochin	0.2 ppm	I	?	?	Effective	Toxic to fish	Schäperclaus, 1954
,, ,,	Aquarium fishes	Plasmoqui n	20 ppm	I 3–20 days	1	....	Effective	....	Van Duijn, 1956
,, ,,	Aquarium fishes	Plasmoquin	10 ppm	I	1	....	Effective	....	Reichenbach-Klinke, 1966
,, ,,	*In vitro*	PAA–2056	2 ppm	I	1	....	Not effective	....	Hoffman and Putz, 1966
,, ,,	Pond fishes	Pine needles	Not given	I	....	....	....	Anti-protozoal action reported	Bauer, 1959
,, ,,	Trout, *Ictalurus punctatus*	PMA	2 ppm	F 1 hr.	....	Every other day	Effective	Toxic to rainbow trout	Foster and Olsen, 1951; Clemens and Sneed, 1958; Snow, 1962
,, ,,	*In vitro*	PMA	0.1 ppm	I	1	....	Effective	Killed trophs and tomites; fish not included	Hoffman and Putz, 1966
,, ,,	Aquarium fishes	Potassium permanganate	4–10 ppm	F 30 min.	....	10 day interval	Effective	Can be toxic in clean water	Van Duijn, 1956

,,	,,	Pond fishes	Potassium permanganate	15 ppm	I	1	....	Effective	In aquaria	R. Allison, 1957a
,,	,,	*In vitro*	Potassium permanganate	2 ppm	I	1	....	Effective	Effectiveness altered by water chemistry, fish not included	Hoffman and Putz, 1966
,,	,,	*Cyprinus carpio*	Quicklime	120 kg/hectare	I	1	As needed	Lowered incidence	Raised pH to 8.2	Ivasik and Karpenko, 1965
,,	,,	*Cyprinus carpio*	Quicklime	3,000 kg/hectare	On drained pond ?	1	....	Lowered incidence	To raise pH	Rychlicki, 1966
,,	,,	Pond fishes	Quinine hydro-chloride	20 ppm	I	3–10	Daily	Effective	....	Gopalakrishnan, 1963 and 1964
,,	,,	*Cyprinus carpio*	Quinine hydro-chloride	10 ppm	I	1	....	....	Effective for ''Ich'' but did not kill *Trichodina*	Schäperclaus, 1954
,,	,,	Aquarium fishes	Quinine hydro-chloride	10 ppm	I	1	As needed	Effective	....	Reichenbach-Klinke, 1966
,,	,,	*In vitro*	Quinine hydrochloride	1 ppm	I	1	....	Effective	Killed trophs and tomites; fish not included	Hoffman and Putz, 1966
,,	,,	*In vitro*	Quinine sulfate	10 ppm	I	1	....	Effective	Killed trophs in 7 hrs. at 20°C	Schäperclaus, 1954
,,	,,	Aquarium fishes	Quinine sulfate or hydrochloride	6.4 ppm	I	2	Daily	Effective	....	Loader, 1963
,,	,,	Aquarium fishes	Quinine sulfate	10 ppm	I	1 or more	As needed	Effective	....	Reichenbach-Klinke, 1966
,,	,,	Pond fishes	Quinine sulfate	8.6 ppm	I	1 or more	As needed	Effective	....	Peterson, *et al*, 1966
,,	,,	*Lebistes reticulatus*	Rivanol	10 ppm	I	?	?	Effective	....	Schäperclaus, 1954

[*continued*

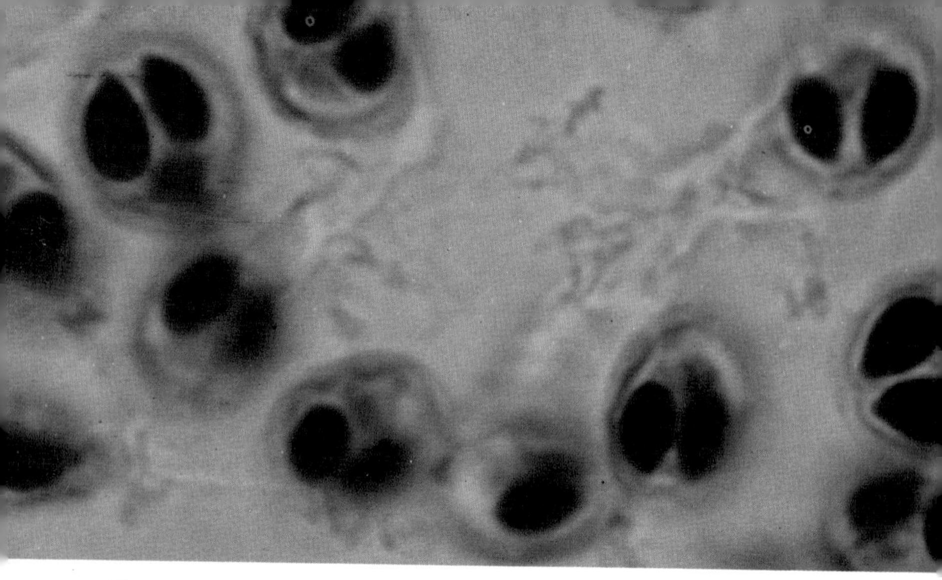

Whirling disease spores in histological section. Polar capsules dark blue, sporoplasm lighter blue. Giemsa stain. × 1000. Preparation and photo by Dr. E. Wood.

Whirling disease. Cross section, 67 days post-infection. Note red-stained trophozoites in normal cartilage, modified Cason Mallory-Heidenhain triple stain. × 200. Experimental infection and photo by J. O'Grodnick and M. Kobularik, Pennsylvania Fish Commission.

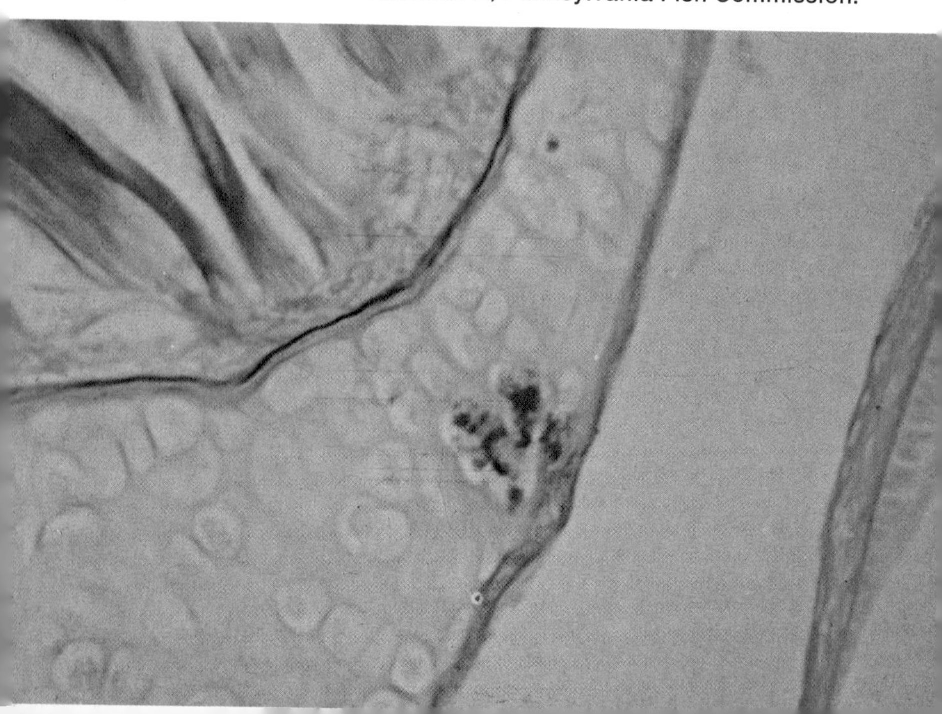

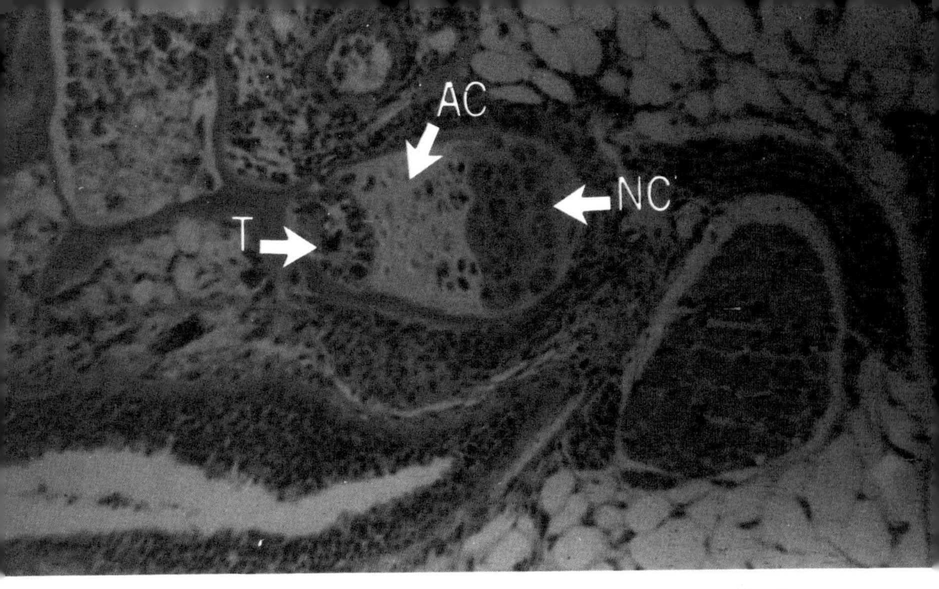

Whirling disease. Cross section of head, 57 days post-infection. Note trophozoites (T), altered cartilage (AC) and normal cartilage (NC). Hematoxylin and eosin. × 100. Photo by G. Hoffman.

Whirling disease. Cross section, 3–4 months post-infection. Note "fairy ring" of red stained trophozoites (T) and adjacent altered cartilage (AC). Modified Cason Mallory-Heidenhain triple stain. × 100. Experimental infection and section by J. O'Grodnick and M. Kobularik, Pennsylvania Fish Commission. Photo by G. Hoffman.

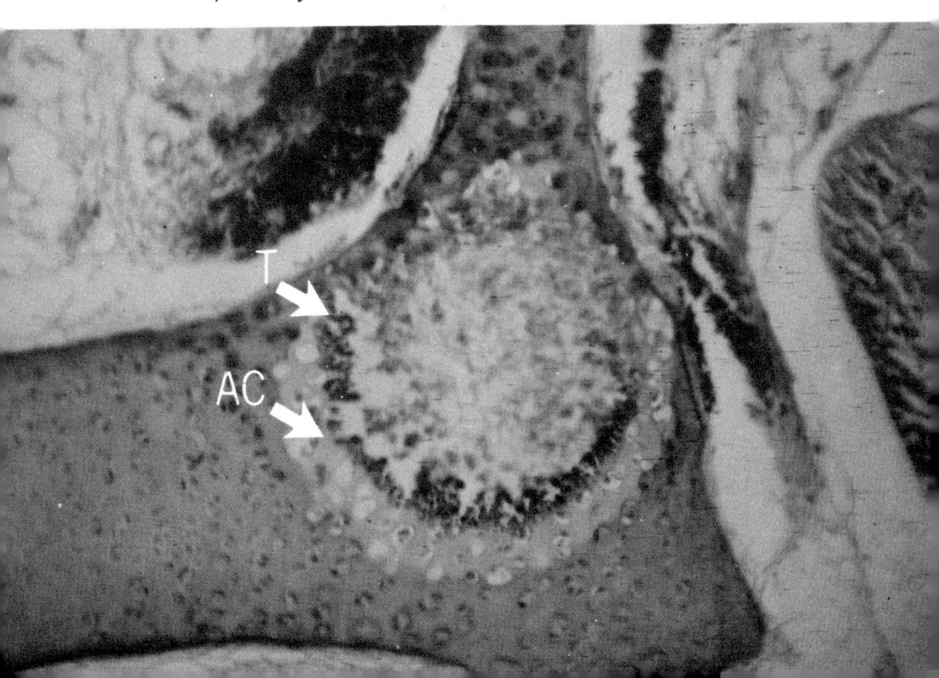

Table 3. Protozoans (External)—*continued*

Parasite	Host	Treatment	Dosage	Method	Number of Applications	Frequency	Author's Report of Success	Remarks	References
Ichthyophthirius multifiliis	*Ictalurus punctatus*	Ruelene	1 ppm	I	1	....	Not effective	....	R. Allison, 1969
,, ,,	*In vitro*	Roccal	0.5 ppm	I	1	....	Not effective	Fish not included	Hoffman and Putz, 1966
,, ,,	Pond fishes	Sodium chloride	7,000 ppm build up to 15,000 ppm	I	1	....	Effective	Leave fish in until cured	Roth, 1922; Schäperclaus, 1954
,, ,,	Trout	Sodium chloride	20,000 ppm	F 1 hr.	As needed	Daily	Effective	....	Butcher, 1947; Gopalakrishnan, 1963 and 1964
,, ,,	Pond fishes	Sodium chloride	30,000 ppm	F 1 hr.	7	Daily	Effective	....	Gopalakrishnan, 1966
,, ,,	Aquarium fishes	Sodium chloride	30,000 ppm	D or F 15–30 min.	As needed	Daily	Effective	Must be repeated	Van Duijn, 1957
,, ,,	Pond fishes	Sodium chloride	7,000 ppm	I	As needed	....	Effective	....	Bauer, 1959
,, ,,	Trout	Sodium chloride	50,000 ppm	D 2–10 min.	As needed	Daily	Beneficial	Disinfect pond, *etc.*	Leitritz, 1960
,, ,,	Pond fishes	Sodium chloride	2,000 ppm	I few days	1	....	Effective	Also controls *Chilodonella*, *Trichodina*	Ivasik and Svirepo, 1964
,, ,,	*Ictalurus punctatus*	Sodium chloride (Sea water)	1,000 ppm	I	1	....	Effective	....	Allen and Avault, 1970
,, ,,	Trout	Sodium chloride (Sea water)	6,500 ppm	I 2 mo.	1	....	Effective	Sea water gradually increased to 0.65%	Dempster, 1970a

,,	,,	Aquarium fishes	Sodium chloride + Potassium permanganate	7,000 ppm 4–4.5 ppm	I	1	....	Effective	....	Van Duijn, 1967
,,	,,	Aquarium fishes	Sodium chlorite ($NaClO_2$)	2–5 ppm	I	....	....	....	Is being tested; shows promise	Dempster, 1970b; Garibaldi, 1971
,,	,,	*In vitro*	Sodium hydroxide	10 ppm	I	1	....	Not effective	....	Hoffman and Putz, 1966
,,	,,	Aquarium fishes	Sulfamezathine, sodium	10 ppm	I	2 or 3	Weekly	Effective	One application usually effective	Postema, 1956
,,	,,	*Ictalurus punctatus*	Sulquin	10,000 ppm in diet	....	....	Daily	Unknown	Sick fish refused to eat	F. Meyer, Unpubl.
,,	,,	*Ictalurus punctatus*	Sulquin	50 ppm	I	....	Daily	Not effective	....	F. Meyer, Unpubl.
,,	,,	Tropical fishes	Temperature, elevated	to 90°F (32°C)	P	1	Daily for 5 days	Effective	Drop slowly at end of 5 days	Stolk, 1956; Schäperclaus, 1954
,,	,,	Tropical fishes	Temperature, elevated	to 87°F (31°C)	P	1	....	Not effective	32°C killed fish	Thadially, 1964
,,	,,	*In vitro*	Tiguvon	0.1 ppm	I	1	1	Not effective	....	Hoffman and Putz, 1966
,,	,,	*Cyprinus carpio*	Transfer method	....	....	As needed	Twice daily	Effective	Transfer fish to clean aquaria	Bauer and Strelkov, 1959
,,	,,	Unidentified	Trypaflavine	10 ppm	I	1	....	Effective	....	Schäperclaus, 1954
,,	,,	*In vitro*	TV 1096	0.2 ppm	I	1	....	Not effective	....	Hoffman and Putz, 1966

[*continued*

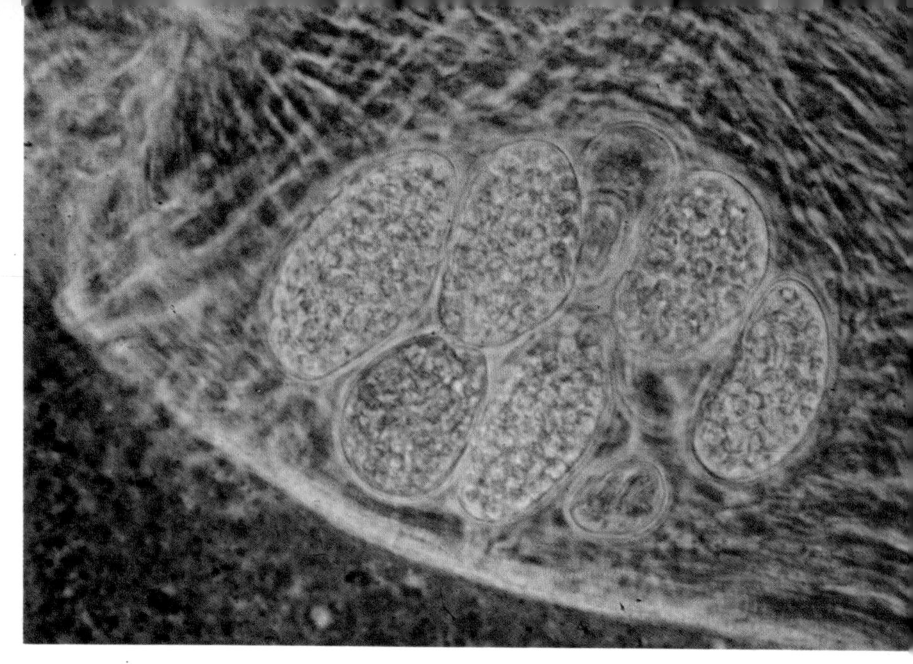

Pleistophora hyphessobryconis sporocysts from *Hyphessobrycon innesi*. Photo by Frickhinger.

Pleistophora hyphessobryconis, free spores. Phase preparation. Note vacuoles in spores. Photo by Frickhinger.

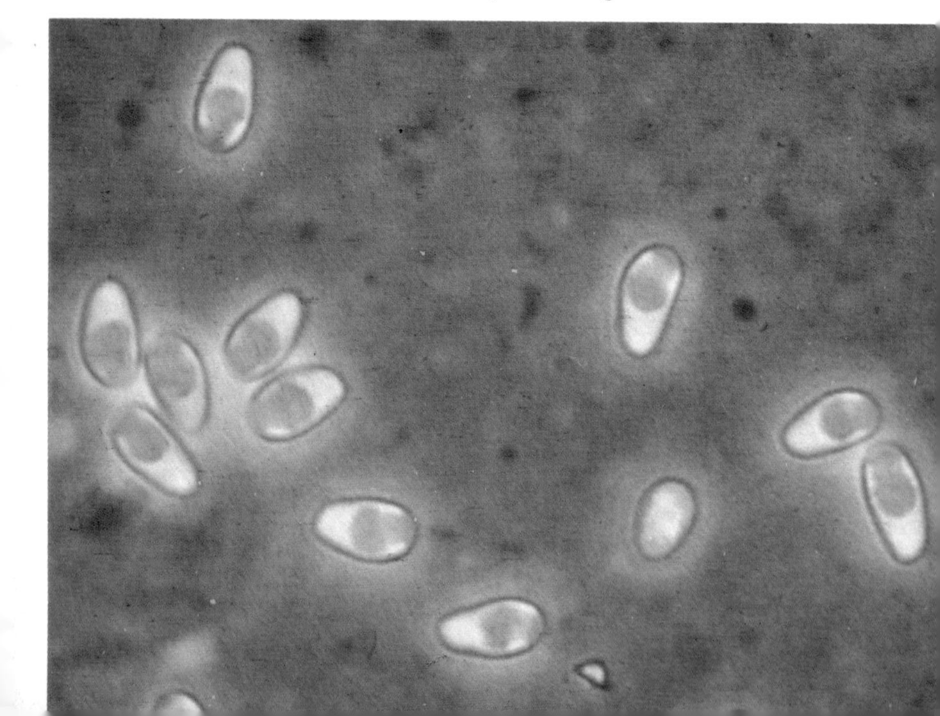

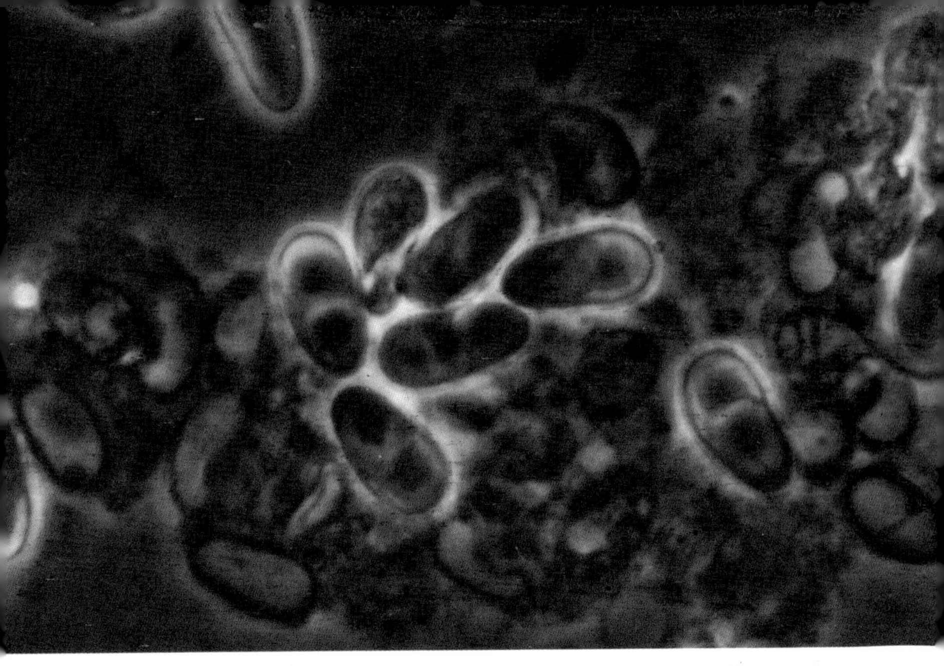

Pleistophora ovariae, spores from ovary of golden shiner. Courtesy of Dr. R. Summerfelt, USFWS, Oklahoma State University.

Cleidodiscus sp. from gills of channel catfish. Note 4 black eyespots at anterior end, and hooks at posterior end. Photo by F. Meyer.

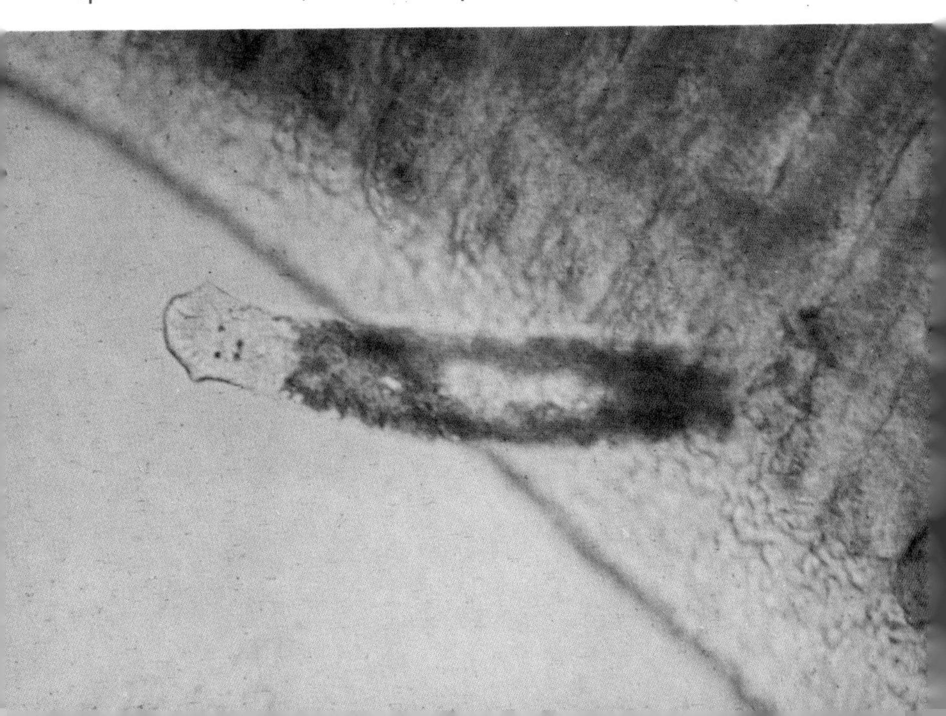

Table 3. Protozoans (External)—*continued*

Parasite	Host	Treatment	Dosage	Method	Number of Appli-cations	Frequency	Author's Report of Success	Remarks	References
Ichthyophthirius multifiliis	*In vitro*	Ultrasonic vibrations	30 khz/70 sec.	P	1	....	Effective	60 sec. killed trophs, 70 sec. killed tomites; fish not included	Nechaeva, 1959
,, ,,	*In vitro*	Ultraviolet light	336,000 MWS per cm^2	P	1	....	Effective	For tomites but not trophs, fish not included	Vlasenko, 1969
,, ,,	*In vitro*	Ultraviolet light	1,717,200 MWS	P	1	....	Effective	For trophs and tomites	Vlasenko, 1969
,, ,,	*In vitro*	Ultraviolet light	90,000 MWS	P	1	....	Effective	Tomites; fish not included	Hoffman, 1970
,, ,,	*In vitro*	Ultraviolet light	MBU–3 (PRK–7)	P	....	Continuous	Effective	Tomites	Kokhanskaya, 1970
,, ,,	*In vitro*	Strong light	?	P	1	....	....	Reduced tomites	Bauer, 1959
,, ,,	Pond fishes	Violet K	0.25–0.5 ppm	I	1	....	Effective	Cheaper than malachite green	Musselius and Filippova, 1968
Oodinium sp.	Unidentified aquarium fishes	Acriflavine	0.2–0.4 ppm	I	?	As needed	Effective	Reduced fecundity of *Lebistes reticulatus*	Patterson, 1950
,, ,,	Aquarium fishes	Acriflavine, neutral	0.6 ppm	I	2	2 day interval	Effective	....	Loader, 1963
,, ,,	Unidentified	Aureomycin	13 ppm	I	?	....	Effective	....	Amlacher, 1961b; Reichenbach-Klinke, 1966
,, ,,	Unidentified	Copper sulfate*	1.5 ppm	I	?	As needed	Effective	Do not use in low carbonate waters	Amlacher, 1961b

*Biological implications of the use of copper sulfate are discussed by Moyle (1949) and Nichols *et al.*, (1946)

,, ,,	Aquarium fishes	Copper sulfate	0.5 ppm†	I	?	As needed	Controlled	Marine fish	Dempster, 1955; Braker, 1961; Høgaard, 1962
,, ,,	Marine aquarium fishes	Copper sulfate pentahydrate 3 parts Citric acid monohydrate 2 parts	0.15 ppm copper by water assay	Continue 10 days	As needed	As needed	Excellent	Also controls other ecto-protozoans	Dempster and Shipman, 1969
Scyphidia sp. (See *Ambiphrya*)	....	....	....	....	....	....	....	....	....
Trichodina indica	Pond fishes	Acetic acid	1,000 ppm	?	1	....	Effective	....	Tripathi, 1954
,, ,,	*Cirrhina reba; Catla* sp.	Formalin	200 ppm	?	1	....	Effective	....	Tripathi, 1954
,, ,,	Pond fishes	Sodium chloride	30,000 ppm	D 5–10 min.	1	....	Effective	....	Tripathi, 1954
Trichodina sp. (See also *Cyclochaeta*)	Trout	Acetic acid, glacial	1,500 ppm	D 1 min	....	As needed	Effective	....	Van Roekel, 1929
,, ,,	*Ictalurus punctatus*	Acriflavine	20 ppm	I	1	....	Inhibitory	....	F. Meyer, Unpubl.
,, ,,	*Ictalurus punctatus, Notemigonus crysoleucas*	Antimycin...A	0.005 ppm	I	2	Weekly	Not effective	....	F. Meyer, Unpubl.
,, ,,	*Cyprinus carpio*	Chloramine-B	20 ppm	I	1	....	Effective	....	Goncharov, 1966
,, ,,	*Notemigonus crysoleucas*	Copper sulfate	0.5–1.0 ppm	I	1	....	Inhibitory	Do not use in low carbonate waters	F. Meyer, Unpubl.
,, ,,	*Ictalurus punctatus*	Dylox	0.25 ppm	I	1	....	Not effective	....	F. Meyer, 1970
,, ,,	Trout	Formalin	166 ppm	F 1 hr.	....	As needed	Effective	Alternate weeks	Fish, 1940

†Concentration is adjusted periodically after determining free copper present. For methods, see Harris (1960)

[*continued*

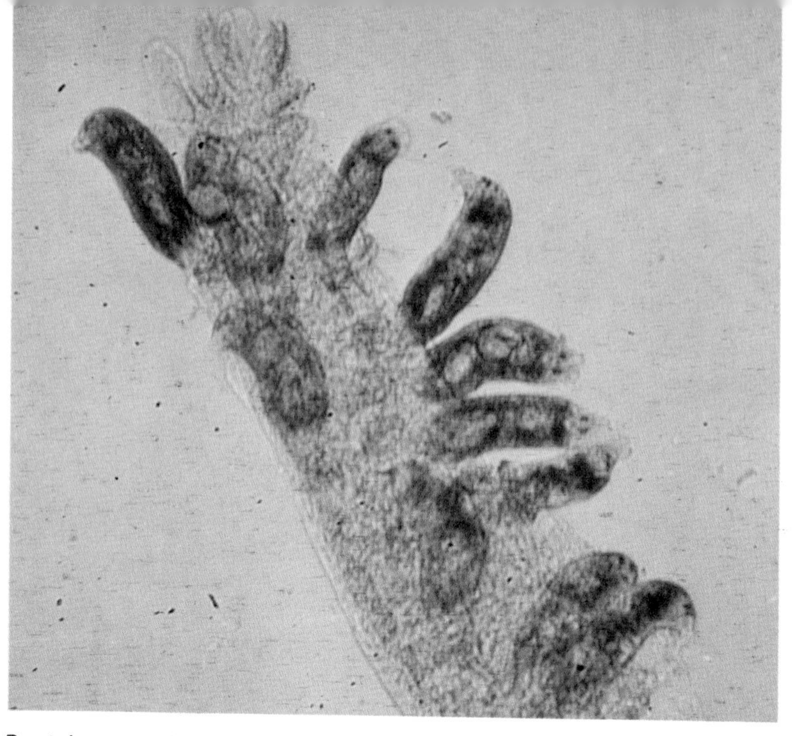

Dactylogyrus juliae on gills of silverjaw minnow. Courtesy of Dr. W. Rogers, Auburn University.

Dactylogyrus sp. eggs may be destroyed by drying. Photo by Frickhinger.

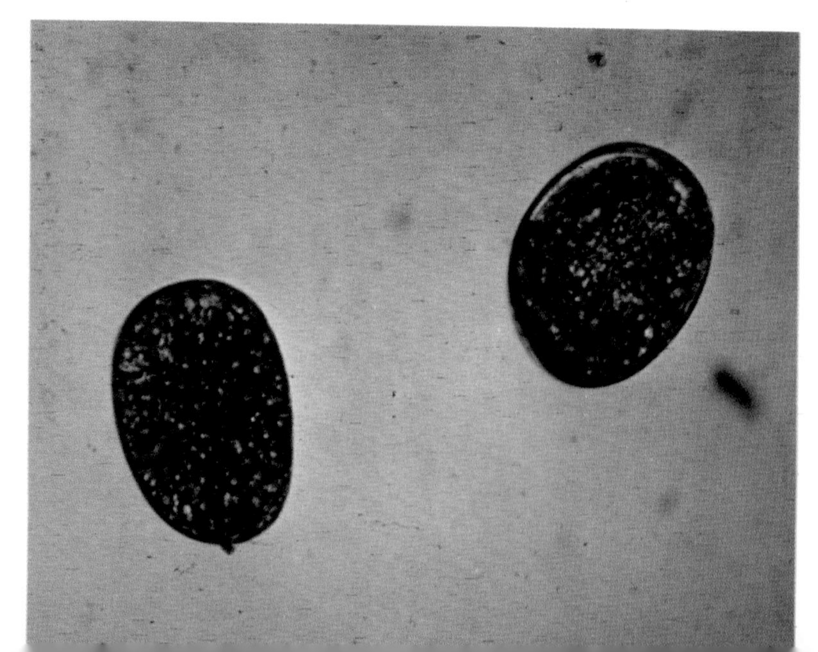

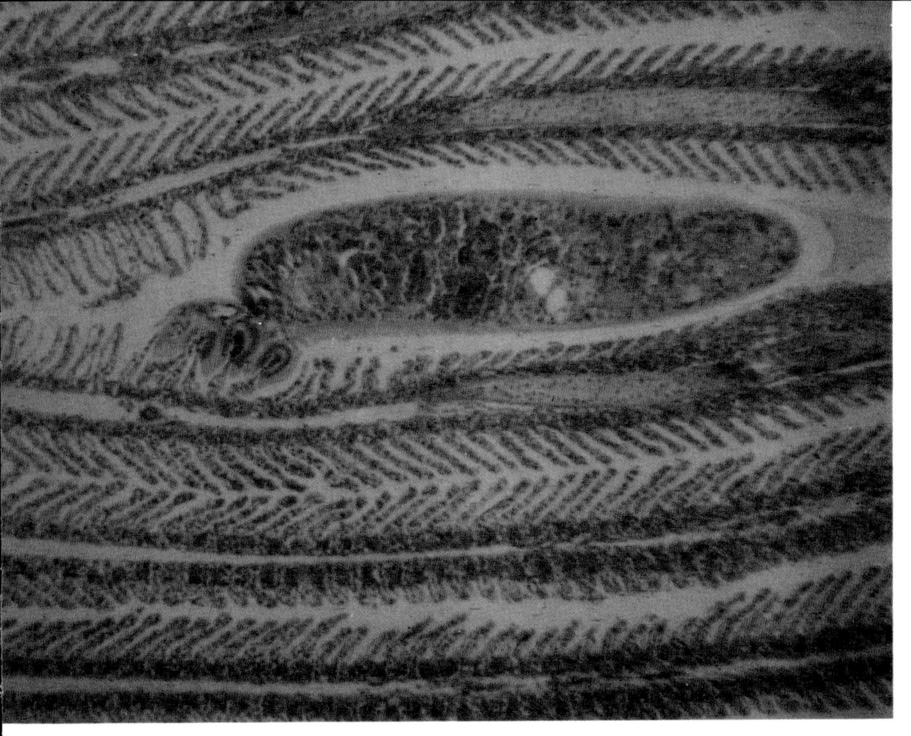

Discocotyle salmonis, gills of Atlantic salmon. Histological section showing haptoral clamps. Preparation by T. Needham, Unilever Research Lab, Aberdeen, Scotland. Photo by G. Hoffman.

Gyrodactyliasis, trout, skin damage and fin necrosis.

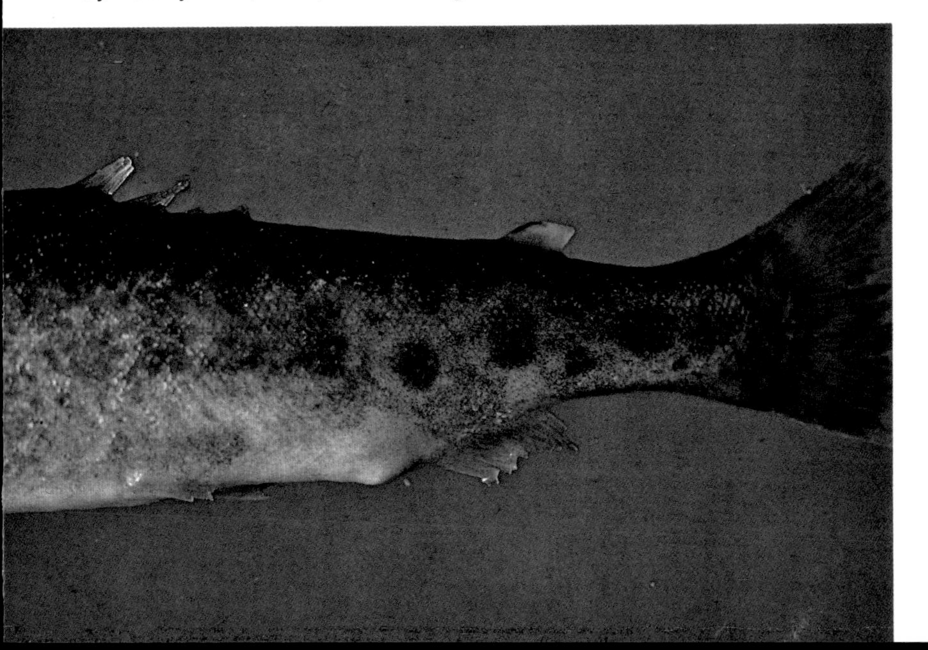

Table 3. Protozoans (External)—*continued*

Parasite	Host	Treatment	Dosage	Method	Number of Applications	Frequency	Author's Report of Success	Remarks	References
Trichodina sp. (See also *Cyclochaeta*)	Salmonids	Formalin	250 ppm	F 1 hr.	1	....	Effective	Use less in warm water	Fish and Burrows, 1940; Davis, 1953
,, ,,	Salmonids	Formalin	166 ppm	F 2 hrs.	1	....	Effective	Recirculated	Fish and Burrows, 1940
,, ,,	Unidentified	Formalin	200 ppm	D 4–8 min.	1	....	Effective	....	Schäperclaus, 1954
,, ,,	*Oncorhynchus tschawytscha*	Formalin	212 ppm	F 1 hr.	1	....	Effective	....	H. Johnson, 1956
,, ,,	Pond fishes	Formalin	15–22 ppm	I	1	....	Effective	Do not use during hot weather	R. Allison, 1957a and 1963; Snow, 1962
,, ,,	Young salmon	Formalin	250 ppm	F 10–15 min.	1	....	Effective	....	Bogdanova, 1962
,, ,,	*Ictalurus punctatus*	Formalin	25 ppm	I	1	....	Effective	Do not use during hot weather	F. Meyer, 1967
,, ,,	Unidentified	Globucid	2,000 ppm	F 24 hrs.	1	....	Effective	....	Schäperclaus, 1954
,, ,,	*Ictalurus punctatus*	Iodoform	2 ppm	I	1	....	Not effective	....	F. Meyer, Unpubl.
,, ,,	*Ictalurus punctatus*	Lysol	200 ppm	D 30 sec.	1	....	Effective	Not as good as formalin	Schäperclaus, 1954
,, ,,	*Ictalurus punctatus*	Malachite green oxalate	0.1 ppm	I	1	....	Effective	....	F. Meyer, Unpubl.
,, ,,	Pond fishes	Malachite green	15 ppm	F	1	6–24 hrs.	Effective	....	Amlacher, 1961a
,, ,,	*Ictalurus punctatus*	Methylene blue	5 ppm	I	1	....	Inhibitory	....	F. Meyer, Unpubl.
,, ,,	*Oncorhynchus tschawytscha*	PMA	3 ppm	F 1 hr.	1	....	Effective	Toxic to *Salmo gairdneri*	H. Johnson, 1956

,,	,,	*Ictalurus punctatus*	PMA	2 ppm	F 1 hr.	1	....	Effective	....	Clemens and Sneed, 1959
,,	,,	*Ictalurus punctatus*	Potassium dichromate	20 ppm	I	1	....	Inhibitory	....	F. Meyer, Unpubl.
,,	,,	*Carassius auratus*	Potassium dichromate	10 ppm	?	?	As needed	Controlled	....	Meehean, 1937
,,	,,	Unidentified	Potassium permanganate	10 ppm	F 1 hr.	?	?	Effective	Toxic to *Stizostedion* sp.	Fish, 1933
,,	,,	*Ictalurus punctatus*	Potassium permanganate	4 ppm	I	1	....	Effective	High organic matter reduces effectiveness	R. Allison, 1957a
,,	,,	Unidentified	Potassium permanganate	10 ppm	F 90 min.	?	Daily	Effective	....	Amlacher, 1961b
,,	,,	*Cyprinus carpio*	Potassium permanganate	1,000 ppm	D 30–45 sec.	?	Daily	Effective	....	Schäperclaus, 1954; Amlacher, 1961b
,,	,,	Pond fishes	Potassium permanganate	1,000 ppm	F 10 min.	As needed	Daily	Effective	....	Amlacher, 1961b
,,	,,	*Ictalurus punctatus, Notemigonus crysoleucas*	Potassium permanganate	2 ppm	I	?	Alternate days	Inhibitory	Organic matter reduces effectiveness	F. Meyer, Unpubl.
,,	,,	*Ictalurus punctatus*	Potassium permanganate	5 ppm	F 16 hrs.	....	....	Effective	Toxic to many species of fish	F. Meyer, Unpubl.
,,	,,	*Ictalurus punctatus*	Potassium permanganate	50 ppm	F 5 min.	1	....	Effective	Toxic to many species of fish	F. Meyer, Unpubl.
,,	,,	*Cyprinus carpio*	Quicklime	2,000 ppm	D 5 sec.	1	....	Effective	....	Schäperclaus, 1954
,,	,,	Unidentified	Quinine hydro-chloride	10 ppm	I	1	....	Not effective	....	Schäperclaus, 1954
,,	,,	*Lebistes reticulatus, Cyprinus carpio*	Roccal	250–500 ppm	F 20–30 min.	1	....	Effective	....	Schäperclaus, 1954

[*continued*

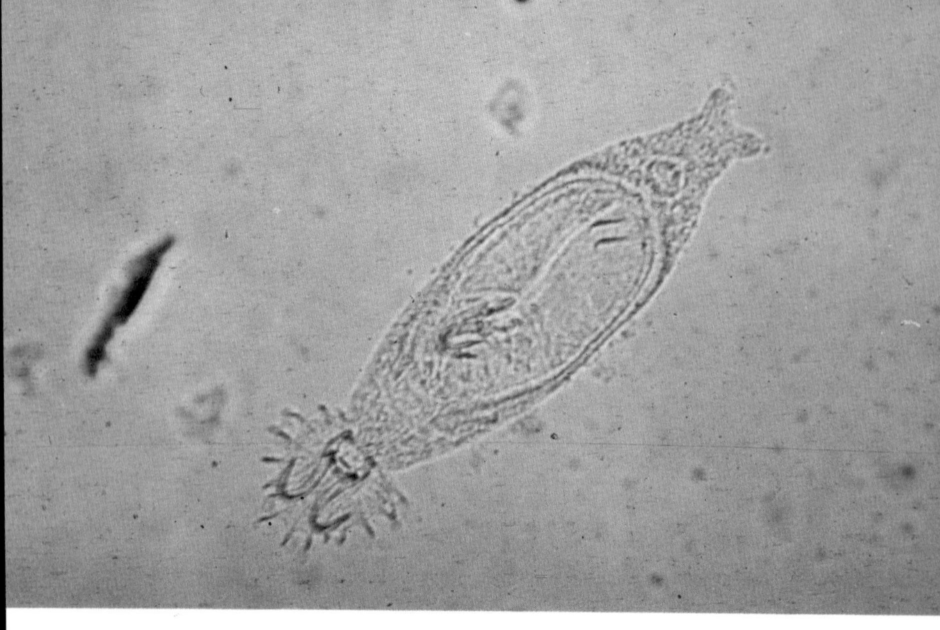

Gyrodactylus sp. Note posterior anchors, embryo, and absence of eye-spots. Photo by F. Meyer.

Clinostomum marginatum, cyst and freed metacercariae. Courtesy of Dr. A. Dechtiar, Fisheries Research Lab, Maple, Ontario.

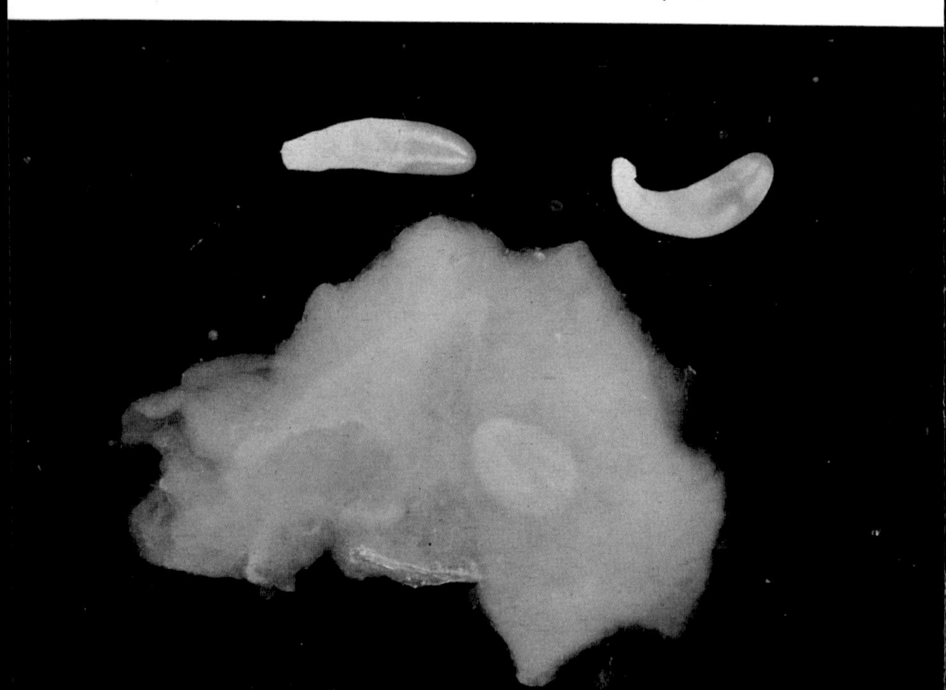

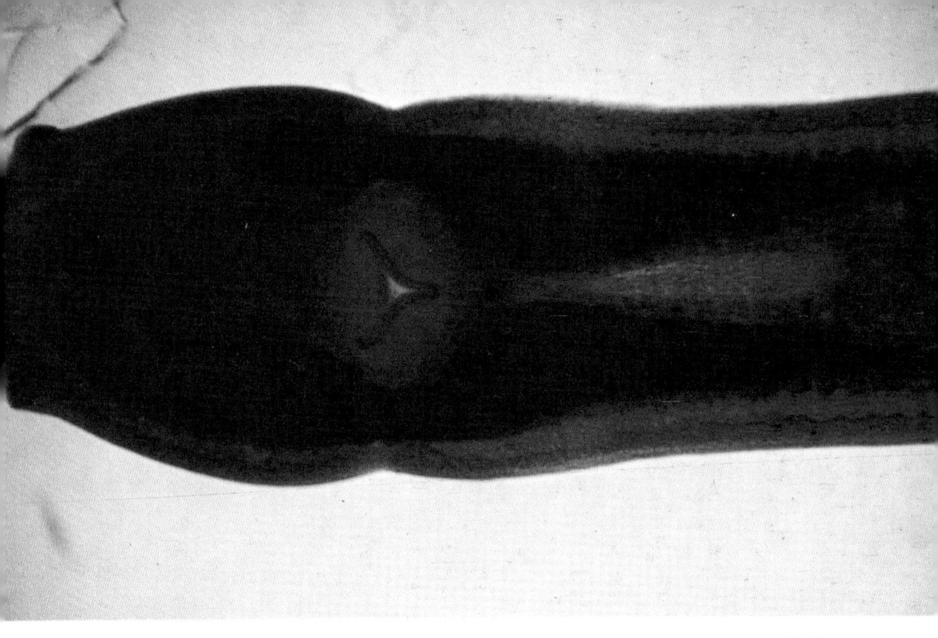

Clinostomum marginatum, anterior end of metacercaria from fish showing "shoulders" around oral sucker and large ventral sucker. Cercariae released from snails infect fish. Photo by G. Hoffman.

Crepidostomum cooperi, stained preparation. Note anterior papillae and stained internal organs. Photo by G. Hoffman.

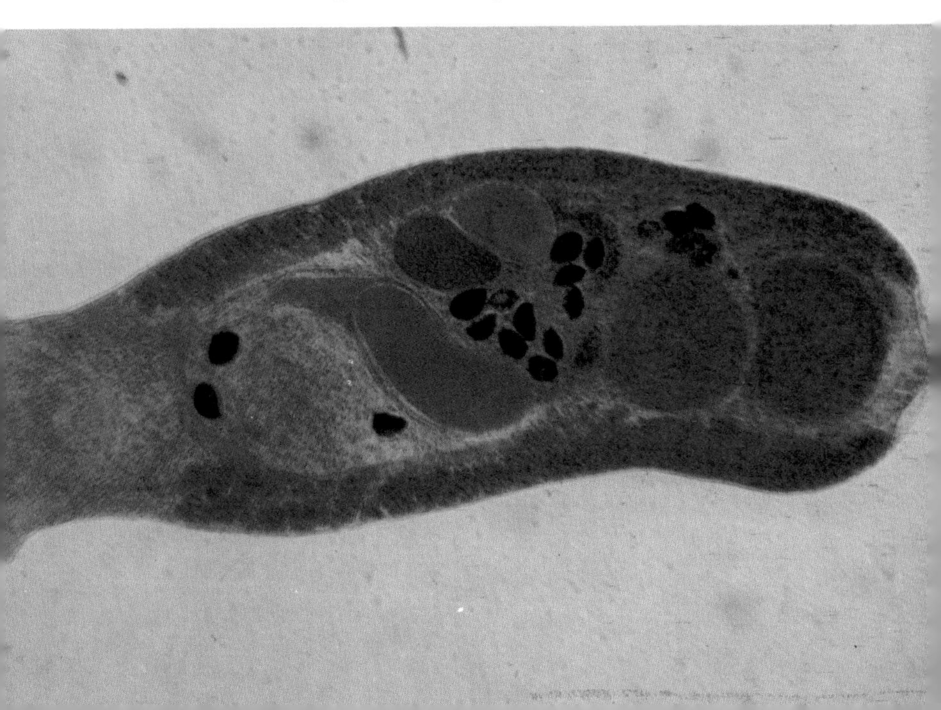

Table 3. Protozoans (External)—*continued*

Parasite	Host	Treatment	Dosage	Method	Number of Appli-cations	Frequency	Author's Report of Success	Remarks	References
Trichodina sp. (See also *Cyclochaeta*)	*Cyprinus carpio*	Sodium chloride	10,000 ppm	D 2–10 min.	As needed	Daily	Effective	Avoid zinc containers	Schäperclaus, 1954
,, ,,	*Cyprinus carpio*	Sodium chloride	17,500 ppm	D 3 min.	As needed	Daily	Effective	Avoid zinc containers	Schäperclaus, 1954
,, ,,	*Cyprinus carpio*	Sodium chloride	25,000 ppm	D 20–25 sec.	As needed	Daily	Effective	Avoid zinc containers	Schäperclaus, 1954
,, ,	Unidentified	Sodium chloride	30,000 ppm	D 5–10 min.	As needed	Daily	Effective	....	Tripathi, 1954
,, ,,	*Salmo salar*	Sodium chloride	30,000 ppm	D 5 min.	As needed	Daily	Effective	....	Bauer and Strelkov, 1959
,, ,,	Unidentified small fish	Sodium chloride	15,000 ppm	F 20 min.	As needed	Daily	Effective	Avoid zinc containers	Amlacher, 1961b
,, ,,	Unidentified large fish	Sodium chloride	25,000 ppm	F 10–15 min.	As needed	Daily	Effective	Avoid zinc containers	Amlacher, 1961b
,, ,,	Pond fishes	Sodium chloride	2,000 ppm	I	1	....	Effective	....	Ivasik and Svirepo, 1964
,, ,,	*In vitro*	Ultraviolet light	160,000–800,000 MWS/cm²	P	1	....	Effective	*Trichodina nigra*; fish not included	Vlasenko, 1969
,, ,,	*In vitro*	Ultraviolet light	70,000 MWS/cm²	P	1	....	Effective	Fish not included	Hoffman, 1971
Trichophrya sp.	Trout	Formalin	250 ppm	F 1 hr.	1	....	Effective	....	Davis, 1953
,, ,,	*Ictalurus punctatus*	Formalin	25 ppm	I	1	....	Not effective	....	F. Meyer, Unpubl.
,, ,,	*Ictalurus punctatus*	Malachite green	0.1 ppm	I	1	....	Inhibitory	....	F. Meyer, Unpubl.
,, ,,	*Ictalurus punctatus*	Malachite green plus formalin	0.1ppm + 25 ppm	I	1	....	Effective	....	F. Meyer, Unpubl.

Unidentified Protozoans	Unidentified	Acriflavine, neutral	Not given	?	?	?	?	Better than acriflavine hydrochloride	Smith, 1942
,, ,,	Pond fishes	Acriflavine	3–5 ppm	F 1–4 hrs.	1	....	Effective	....	Snow, 1962
,, ,,	Unidentified	Calcium hypochorite	100 ppm	?	?	....	Not effective	....	Reichenbach-Klinke, 1966
,, ,,	Unidentified	Chloramine-T	75 ppm	F 1 hr.	1	....	Not effective	....	Fish, 1939
,, ,,	Unidentified	Chloramine-T	166 ppm	F 30 min.	1	....	Not effective	....	Fish, 1939
,, ,,	Unidentified	Chloramine-B	Not given	?	?	....	?	Do not use in metal tanks	Goncharov, 1966
,, ,,	Trout	Formalin	333–500 ppm	F	?	....	Effective	First to use prolonged treatment	Kingsbury and Embody, 1932
,, ,,	Unidentified	Formalin	250 ppm	F 1 hr.	?	As needed	Effective	Toxic to *Dorosoma cepedianum*	Peterson, *et al.*, 1966
,, ,,	*Salmo gairdneri*	Formalin	200 ppm	F 1 hr.	1	....	Not effective	At 3°C	BSFW, Hatchery Biologist's Reports, 1969
,, ,,	*Salmo gairdneri*	Formalin	250 ppm	F 1 hr.	1	....	Effective	At 7°C	BSFW, Hatchery Biologist's Reports, 1969
,, ,,	*Salmo gairdneri*	Formalin	250 ppm	F 1 hr.	1	....	Effective	Toxic at 10°C	BSFW, Hatchery Biologist's Reports, 1969
,, ,,	Unidentified	Hydrogen peroxide	525 ppm	D 10–15 min.	?	As needed	Effective	....	Reichenbach-Klinke, 1966
,, ,,	Salmonids	Malachite green oxalate	1.25–5 ppm	F 30 min.	1	....	Effective	....	Barney, 1963

[continued

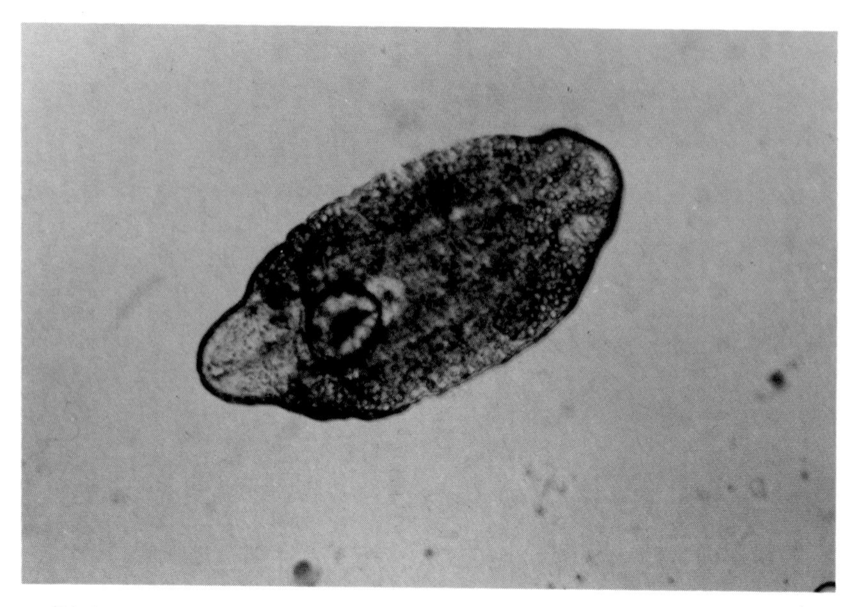

Diplostomum spathaceum metacercaria from the lens of blinded fish. This and other trematodes are transmitted to fish by snails.

Nanophyetus salmincola, cercariae from snails infect fish and cause exopthalmus and kidney damage. Courtesy of Dr. R. Milleman and American Fisheries Society.

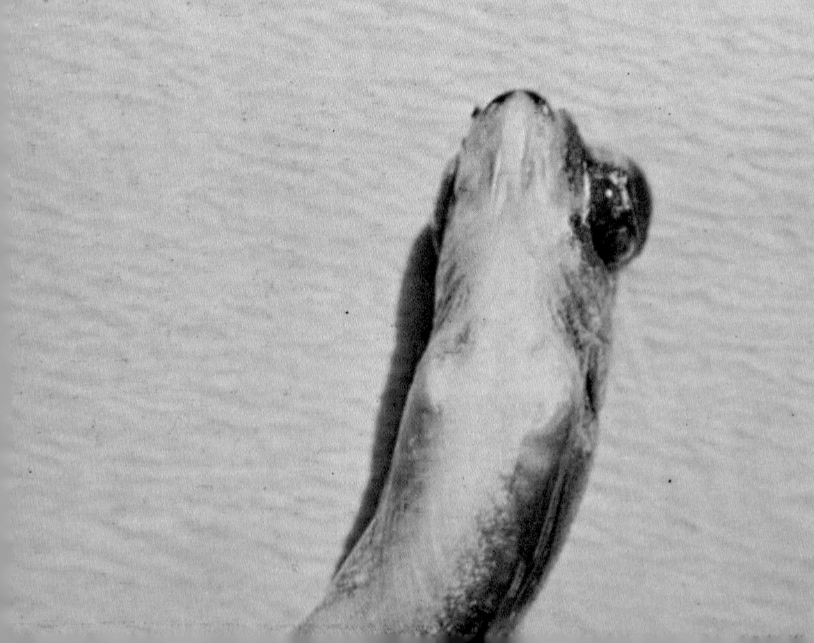

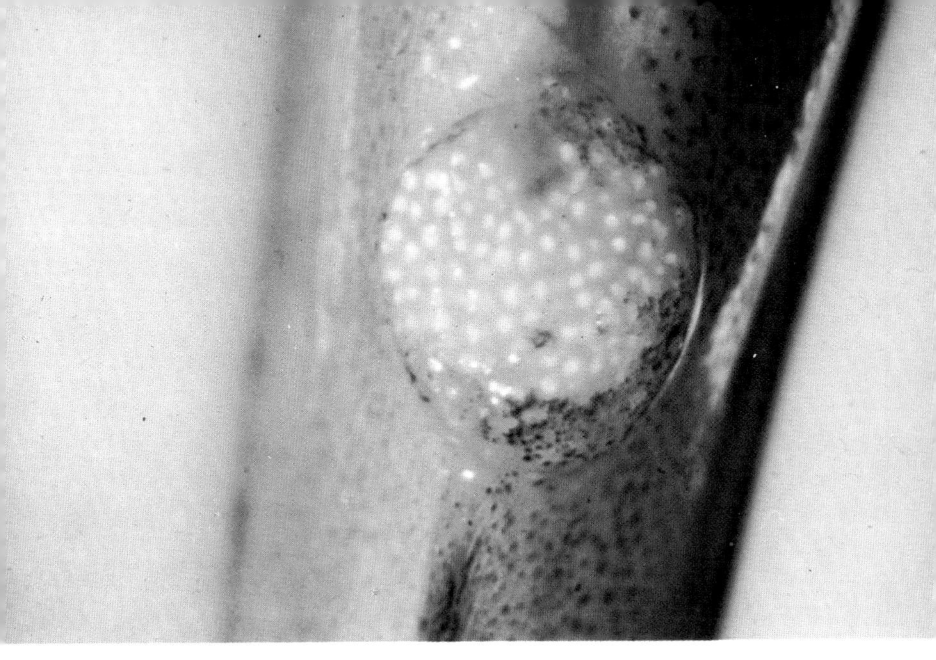

Nanophyetus salmincola metacercarial cysts causing prolapsed intestine of coho salmon. Courtesy of Dr. R. Milleman and American Fisheries Society.

Nanophyetus salmincola metacercariae in a wet squash. Dark oval objects are opaque excretory bladders. Courtesy of Dr. R. Milleman, Oregon State University.

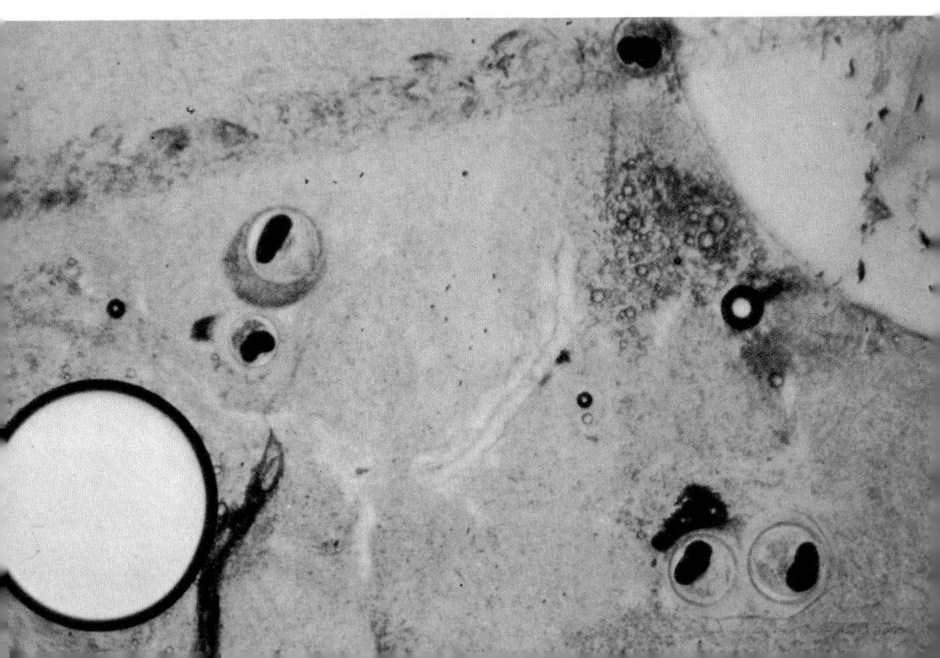

Table 3. Protozoans (External)—*continued*

Parasite	Host	Treatment	Dosage	Method	Number of Applications	Frequency	Author's Report of Success	Remarks	References
Unidentified Protozoans	*Salmo salar*	Malachite green	1.25–5 ppm	F 30 min.	1	....	Effective	....	Dexter, 1963
,, ,,	Aquarium fishes	Methylene blue	2–4 ppm	I	1	....	Effective	Toxic to plants	Van Duijn, 1967
,, ,,	Unidentified	Plasmochin	10 ppm	I	1	....	Effective	....	Reichenbach-Klinke, 1966
,, ,,	Trout	PMA	2 ppm	F 1 hr.	I	....	Effective	Toxic to *Salmo gairdneri*	Foster and Olson, 1951
,, ,,	Unidentified	Potassium dichromate	5 ppm	I	1	....	Effective	....	Reichenbach-Klinke, 1966
,, ,,	Trout	Pottasium permanganate	3.3 ppm	F 30 min.	?	Weekly	Effective	High organic matter reduces effectiveness	Prevost, 1934
,, ,,	*Dorosoma cepedianum*	Potassium permanganate	3 ppm	F 1 hr.	?	As needed	Effective	....	Peterson, *et al.*, 1966
,, ,,	Unidentified	Rivanol	2–5 ppm	I	1	....	Effective	....	Amlacher, 1961b; Reichenbach-Klinke, 1966

Table 4. PROTOZOANS (INTERNAL)

Parasite	Host	Treatment	Dosage	Method	Number of Applications	Frequency	Author's Report of Success	Remarks	References
Ceratomyxa shasta	*Salmo gairdneri* other salmonids	Chlorine	0.3 ppm	I	Continuous	....	Effective	Water had to be dechlorinated with activated carbon	Bedell, 1971; Leith and Moore, 1967
,, ,,	*Salmo gairdneri*	Filtration	....	I	Continuous	....	Partially effective	Infection rate reduced	Leith and Moore, 1967
,, ,,	Salmonids	Ultraviolet light	37,000 MWS/cm²	I	Continuous	....	Effective	Possibly excessive	Burrows, 1971
,, ,,	*Salmo gairdneri*	Ultraviolet light	G 36T–66 lamps	I	Continuous	....	Effective	....	Bedell, 1971; Leith and Moore, 1967
,, ,,	*Salmo gairdneri*	Ultraviolet light	215,560 MWS/cm²	I	Continuous	....	Effective	Probably excesssive	Sanders, *et al.*, 1972
Eimeria sp.	*Ctenopharyngodon idellus*	Calcium chloride	5 centners/hectare	M	1	....	Effective	For disinfecting pond bottom	Musselius and Strelkov, 1968
,,	*Ctenopharyngodon idellus*	Furoxone	Not given	O	?	?	Effective	....	Musselius and Strelkov, 1968
,,	*Ctenopharyngodon idellus*	Furoxone	Not given	O	?	?	Effective	....	Kulow and Spangenberg, 1969
,,	*Ctenopharyngodon idellus*	Stovarsol	1 mg/g of food	O	?	?	Effective	....	Naumova and Kanaev, 1962
Hexamita sp.	Trout	Aureomycin	10,000 ppm in food	O	....	....		....	Van Duijn, 1967
,,	Trout	Betanapthol	5 g/kg of fish	O	4–5	Daily	....	....	Reichenbach-Klinke, 1966

[*continued*

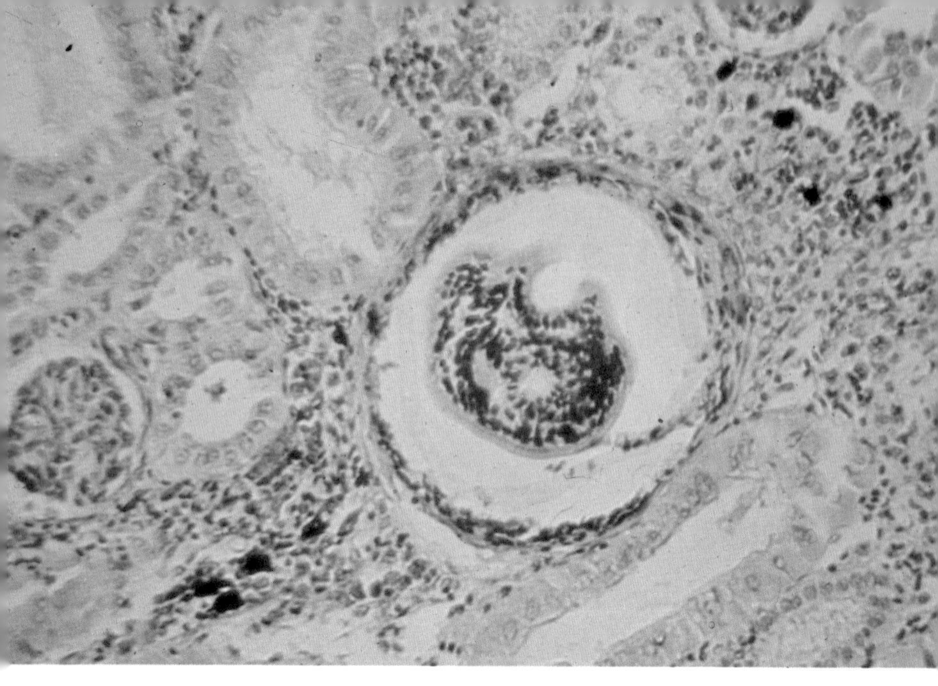

Nanophyetus salmincola. Histologic section of parasite cyst in kidney of salmon. Note cup-shaped sucker. Courtesy of S. Knapp and Dr. R. Milleman, Oregon State University.

Bothriocephalus cuspidatus. Cross section of parasite (P) in cecum (C) of bluegill. Photo by G. Hoffman.

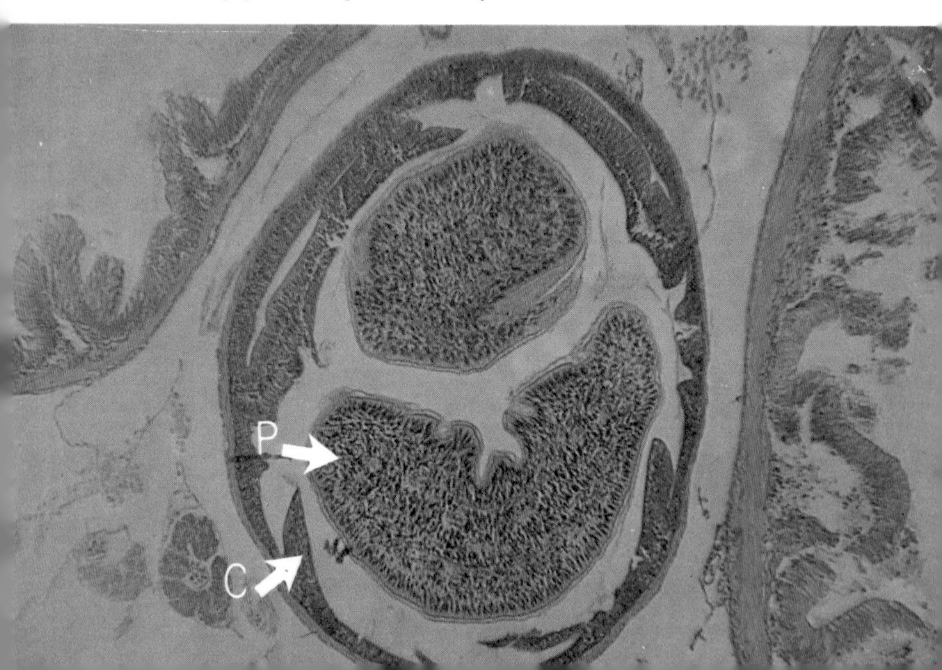

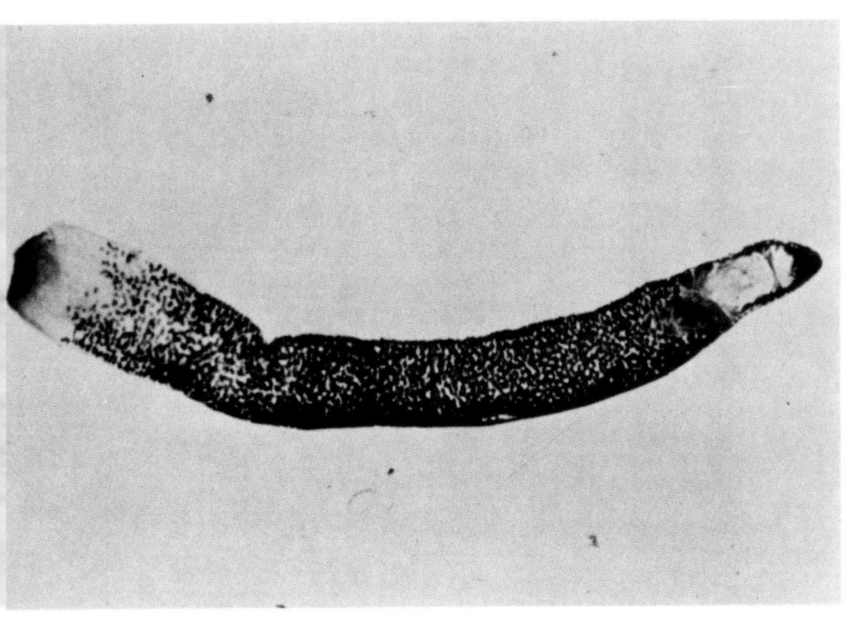

Caryophyllaeides fennica from intestine of European bream.

Corallobothrium sp. in intestine of channel catfish. Courtesy of Dr. W. Rogers, Auburn University.

Table 4. Protozoans (Internal)—*continued*

Parasites	Host	Treatment	Dosage	Method	Number of Applications	Frequency	Author's Report of Success	Remarks	References
Hexamita sp.	Trout	Calomel	2,000 ppm in food	0	4	Daily	Effective	....	Smith and Quistorff, 1940; Reichenbach-Klinke, 1966
,,	Trout	Carbarsone (Arthinol)	2,000 ppm in food	0	3	Daily	Effective	....	Nelson, 1941; Yasutake, *et al.*, 1961; Reichenbach-Klinke, 1966
,,	Salmonids	Carbarsone oxide	2,000 ppm in food	0	3	Daily	Effective	....	Yasutake, *et al.*, 1961
,,	Trout	Cyzine	20 ppm in food	0	3	Daily	Effective	....	McElwain and Post, 1968
,,	Salmonids	Dimetridazole	....	....	....	....	....	A good candidate for hexamitiasis	Hoffman, Upubl.
,,	Salmonids	Enheptin	2,000 ppm in dry food	0	3	Daily	Effective	....	Yasutake, *et al.*, 1961; Post and Beck, 1966
,,	Trout	Emtrysidina	15,000 ppm in dry food	0	3	Daily	Effective	Contains dimetridazole and amminosidin	Ghittino, 1968
,,	Trout	Entobex	10,000 ppm in food	0	4	Daily	Effective	....	Van Duijn, 1967
,,	Salmonids	Fumagillin	20,000 ppm in food	0	3	Daily	Effective	Available as Fumadil.©	Yasutake, *et al.*, 1961
,,	Trout	Magnesium sulfate	30,000 ppm in dry food	0	3	Daily	Effective	....	BSFW, Hatchery Biologist's Reports, 1968

,,	Trout	Phenothiazine	Not given	0	3–4	Daily	Effective	May cause "back peel"	H. Wolf *in* Rucker, 1957
,,	Salmonids	PR-3714 (Abbott)	2,000 ppm in dry food	0	3	Daily	Effective	....	Yasutake, *et al.*, 1961
,,	Trout	Stovarsol (Acetarsone)	10 mg/kg of fish	0	3–4	Daily	Effective	....	Reichenbach-Klinke, 1966
Myxosoma cerebralis	*In vitro*	Ammonium carbonate	1,000 ppm	I-M	1	....	Not effective	In 14 days	Hoffman and Hoffman, 1972
,, ,,	*In vitro*	Ammonium chloride	1,000 ppm	I-M	1	....	Not effective	In 14 days	Hoffman and Hoffman, 1972
,, ,,	Salmonids	Calcium cyanamide	2,000 kg/hectare (1,833 lbs/acre)	Apply when wet	1	Annually	Effective	In drained ponds	Schäperclaus, 1932
,, ,,	Salmonids	Calcium cyanamide	5,000 kg/hectare (4,453 lbs/acre)	Apply when wet	2	3–4 months	Effective	In drained ponds	Tack, 1951
,, ,,	Salmonids	Calcium cyanamide	5,000 kg/hectare (4,453 lbs/acre)	Apply when wet	2	3-4 months	Effective	In drained ponds	Ghittino, 1970
,, ,,	*In vitro*	Calcium hydroxide	5,000 ppm	I-M	1	....	Not effective	In 14 days	Hoffman & Hoffman, 1972
,, ,,	Simulated ponds	Calcium hypochlorite	300 ppm (as free chlorine)	M (16 hrs)	1	....	....	10 of 10 test fish infected	Hoffman, 1972
,, ,,	Simulated ponds	Calcium hypochlorite	600 ppm (as free chlorine)	M (16 hrs)	1	....	....	5 of 10 test fish infected	Hoffman, 1972
,, ,,	Simulated ponds	Calcium hypochlorite	1,200 ppm (as free chlorine)	M (16 hrs)	1	....	....	2 of 30 test fish infected	Hoffman, 1972
,, ,,	*In vitro*	Calcium hypochlorite	400 ppm (as free chlorine)	I-M	1	....	Not effective	In 14 days	Hoffman and Hoffman, 1972
,, ,,	*In vitro*	Calcium oxide	2,500 ppm	I-M	1	....	Effective	In 6 days	Hoffman and Hoffman, 1972

[*continued*

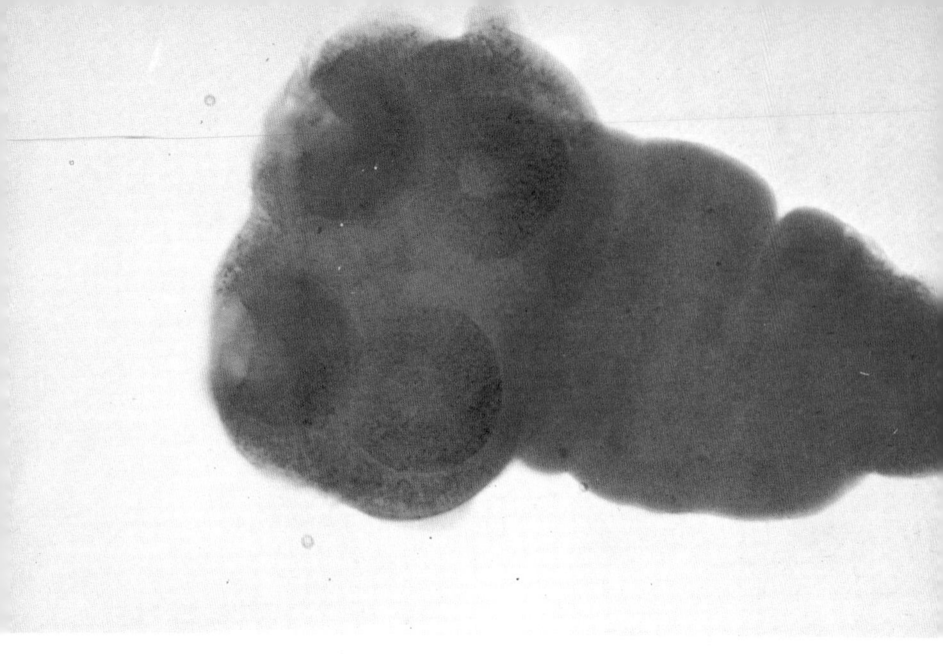

Corallobothrium sp. scolex from channel catfish. Courtesy of K. Sneed, FFES, USF & WS.

Diphyllobothrium sebago, white plerocercoid going through liver of small Atlantic salmon. Courtesy of R. Dexter, USF & WS.

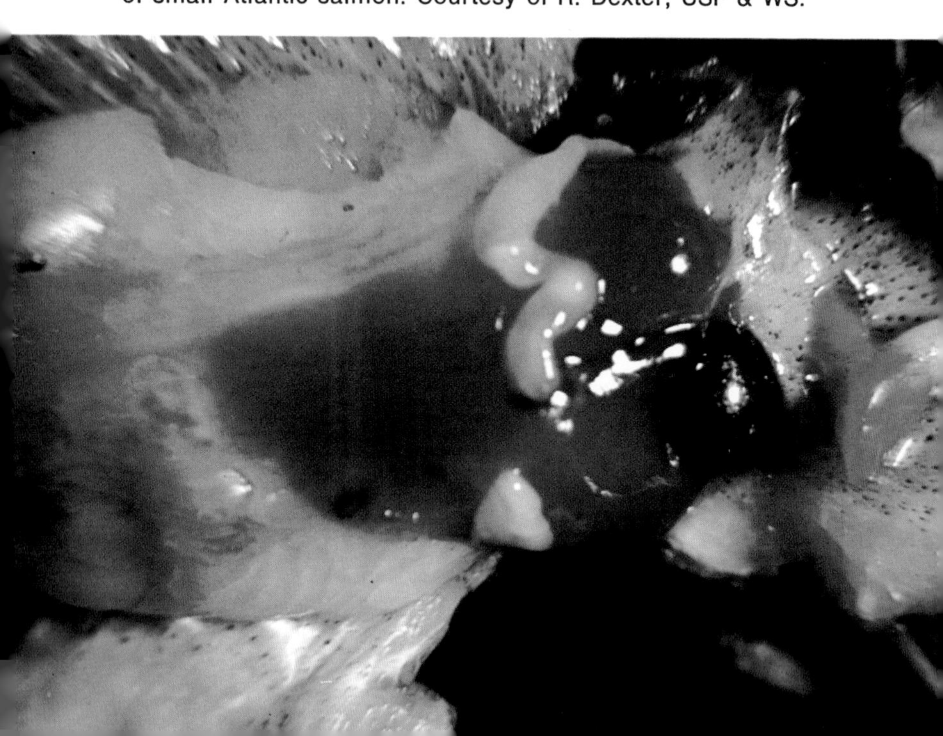

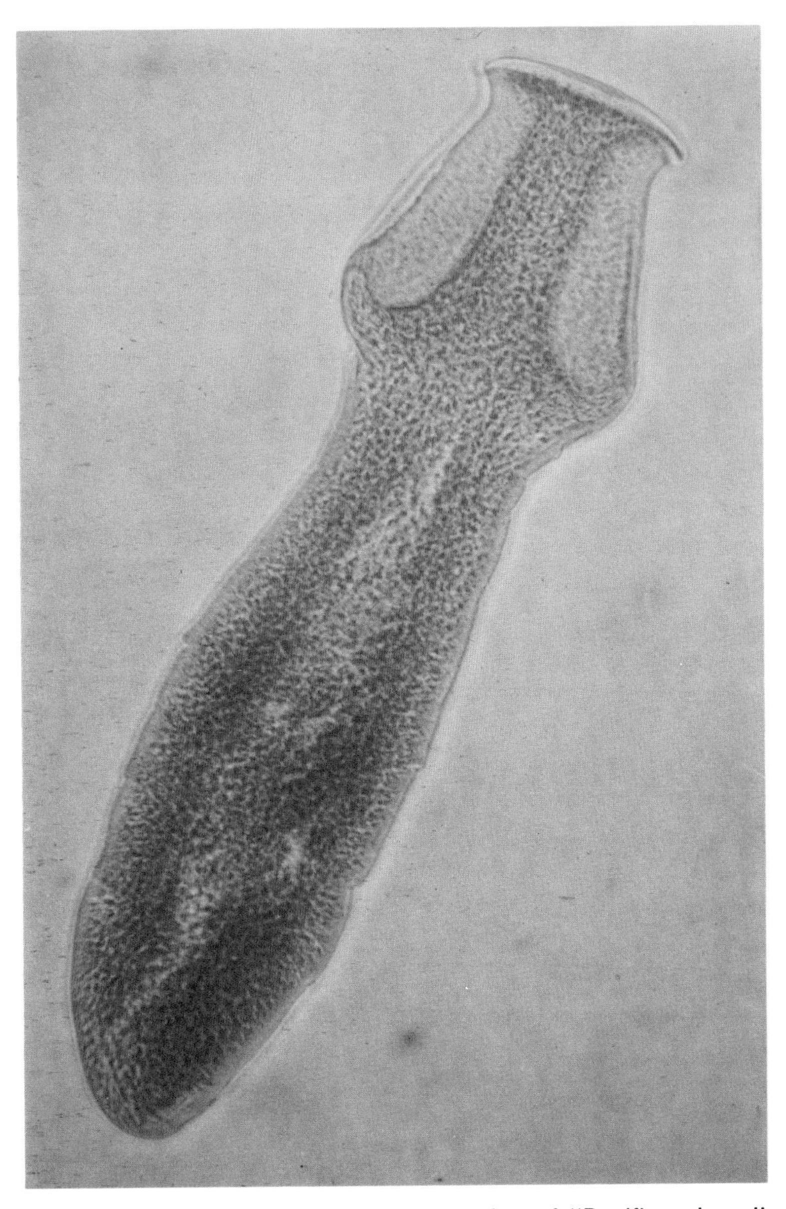

Eubothrium salvelini juvenile from intestine of "Pacific salmon". Courtesy of Dr. L. Margolis, FRBC, Pacific Biological Station, Nanaimo, B.C., Canada.

Table 4. Protozoans (Internal)—*continued*

Parasite	Host	Treatment	Dosage	Method	Number of Appli-cations	Frequency	Author's Report of Success	Remarks	References
Myxosoma cerebralis	*In vitro*	Calcium oxide	5,000 ppm	I-M	1	....	Effective	In 2 days	Hoffman and Hoffman, 1972
,, ,,	*In vitro*	Copper sulfate	5,000 ppm	I-M	1	....	Not effective	In 14 days	Hoffman and Hoffman, 1972
,, ,,	*In vitro*	Potassium hydroxide	1,000 ppm	I-M	1	....	Not effective	In 14 days	Hoffman and Hoffman, 1972
,, ,,	*In vitro*	Potassium hydroxide	10,000 ppm	I	1	....	Effective	In 2 days	Hoffman and Hoffman, 1972
,, ,,	*In vitro*	Potassium permanganate	10,000 ppm	I	1	....	Not effective	In 14 days	Hoffman, and Hoffman, 1972
,, ,,	Salmonids	Quicklime*	Not given	....	....	....	Calcium cyanamide better	In drained ponds	Schäperclaus, 1954
,, ,,	Salmonids	Quicklime	1,250 kg/hectare	....	....	....	Calcium cyanamide better	In drained ponds	Ghittino, 1970
,, ,,	*In vitro*	Roccal	200 ppm	I	1	....	Not effective	In 14 days	Hoffman and Hoffman, 1972
,, ,,	*In vitro*	Sodium borate	1,000 ppm	I	1	....	Not effective	In 14 days	Hoffman and Hoffman, 1972
,, ,,	*In vitro*	Ultraviolet light	35,000 MWS/cm²	I	1	Continuous	Effective	Destroys spores in water supply	Hoffman, Unpubl.

*Snow and Jones (1959) found that hydrated lime could be used to produce as alkaline a pH as quicklime in bluegill pond treatment.

Table 5. COELENTERATES

Parasite	Host	Treatment	Dosage	Method	Number of Applications	Frequency	Author's Report of Success	Remarks	References
Hydra sp.	Unidentified	Ammonium nitrate	50,000 ppm	D	?	?	Effective	....	Reichenbach-Klinke and Elkan, 1965
"	Unidentified	Aquarol	80 ppm	D	?	3rd day	Effective	....	Reichenbach-Klinke and Elkan, 1965

Eubothrium salvelini from intestine of "Pacific salmon". Courtesy of Dr. L. Margolis, FRBC, Pacific Biological Station, Nanaimo, B.C., Canada.

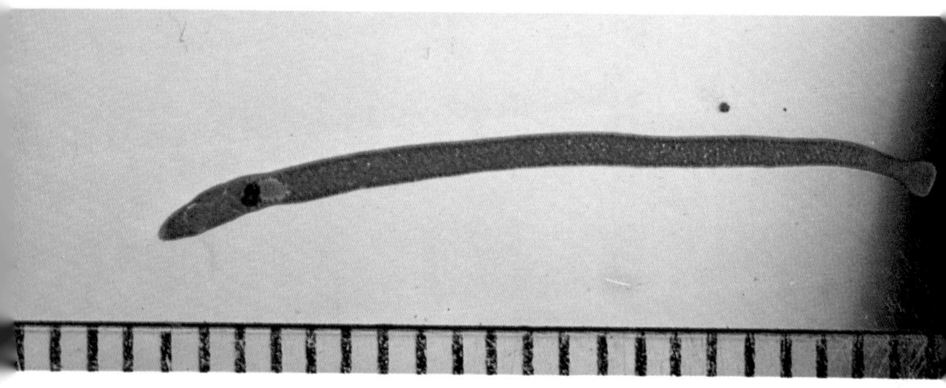

Glaridacris catostomi, a relative of *Caryophyllaeus* (p. 134) from the intestine of *Catostomus commersoni.* Courtesy of Dr. J. S. Mackiewicz, State University of New York at Albany.

Isoglaridacris bulbocirrus, a close relative of *Caryophyllaeus* (p. 134) from the intestine of *Catostomus commersoni*. Courtesy of Dr. J. S. Mackiewicz, State University of New York at Albany.

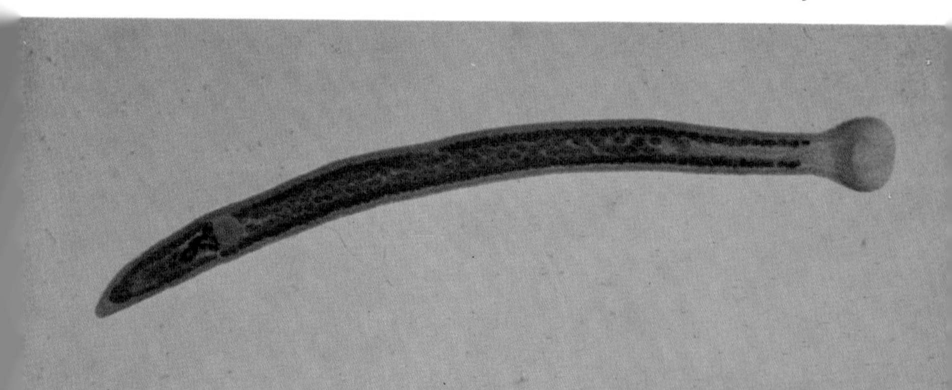

Ligula intestinalis from body cavity of spot-tail shiner. Courtesy of Dr. A. Dechtiarenko, Fisheries Research Lab, Maple, Ontario, Canada.

Proteocephalus ambloplitis plerocercoids have contributed to the emaciation of this adult largemouth bass. Bass with smaller numbers of parasites may appear normal. Courtesy of Dr. C. Johnson, University of North Carolina.

Table 6. MONOGENETIC TREMATODES

Parasite	Host	Treatment	Dosage	Method	Number of Appli-cations	Frequency	Author's Report of Success	Remarks	References
Benedinia seriolae	*Seriola aureovittata*	Sodium perborate ($NaBO_3.4(H_2O)$)	500 ppm	D 8.5 min.	1	....	Effective	....	Hoshina, 1966
,, ,,	*Seriola aureovittata*	Sodium perborate	1,000 ppm	F 30 min.	1	....	Effective	Did not kill eggs of parasite	Hoshina, 1966
Unidentified Monogenea	*Seriola aureovittata*	Sodium perborate	1,000 ppm	D 2–3 min.	1	....	Effective	....	Hoshina 1966
,, ,,	*Seriola aureovittata*	Sodium perborate	10,000 ppm	D 1 min.	1	....	Effective	....	Kasahara, 1967 and 1968
,, ,,	*Seriola aureovittata*	Sodium peroxide pyrophosphate	1,000 ppm	?	1	....	Effective	....	Hoshina, 1966
,, ,,	*Seriola aureovittata*	Sodium peroxide pyrophosphate	10,000 ppm	D 15–30 sec.	1	....	Effective	....	Kasahara, 1967
Cleidodiscus pricei	*Ictalurus punctatus*	Formalin	10–15 ppm	I	1	....	Effective	....	R. Allison, 1963
Cleidodiscus sp.	*Ictalurus punctatus*	Acriflavine	20 ppm	I	1	....	Not effective	....	F. Meyer, Unpubl.
,, ,,	*Ictalurus punctatus*	Antimycin A	0.005 ppm	I		4 Weekly	Not effective	Toxic to small fish	F. Meyer, Unpubl.
,, ,,	*Ictalurus punctatus*	Co-Ral	1 ppm	I	1	....	Effective	....	R. Allison, 1969
,, ,,	*Ictalurus punctatus*	Dylox	0.25 ppm	I	1	....	Effective	....	F. Meyer, 1970
,, ,,	*Ictalurus punctatus*	Formalin	25 ppm	I	1	....	Effective	Recurrences are common	F. Meyer, Unpubl.
,, ,,	*Ictalurus punctatus*	Gentian violet	0.3 ppm	I	1	....	Not effective	Toxic to some lots of fish	F. Meyer, Unpubl.
,, ,,	*Ictalurus punctatus*	Iodoform	2.0 ppm	I	1	....	Not effective	....	F. Meyer, Unpubl.
,, ,,	*Ictalurus punctatus*	Methylene blue	50 ppm	I	1	....	Not effective	....	F. Meyer, Unpubl.

,, ,,	*Ictalurus punctatus*	Potassium dichromate	20 ppm	I	1	....	Not effective	....	F. Meyer, Unpubl.
,, ,,	*Ictalurus punctatus*	Potassium permanganate	2 ppm	I	1	....	Effective	Recurrences are common	F. Meyer, Unpubl.
,, ,,	*Ictalurus punctatus*	Potassium permanganate	5 ppm	F 16 hrs.	1	....	Effective	Toxic to many lots of fish	F. Meyer, Unpubl.
,, ,,	*Ictalurus punctatus*	Potassium permanganate	50 ppm	D 5 min.	1	....	Effective	Toxic to many lots of fish	F. Meyer, Unpubl.
,, ,,	*Ictalurus punctatus*	Ruelene	0.5 3 ppm	I	1	....	Effective	....	R. Allison, 1969
Dactylogyrus anchoratus	*Cyprinus carpio*	Quinine hydrochloride	20 ppm	F 6 hrs.	1	....	Not effective	Toxic to fish	Schäperclaus 1954
Dactylogyrus extensus	*Cyprinus carpio*	Bromex-50	0.2 ppm	I	1	....	Effective	....	Lahav, *et al.*, 1966
,, ,,	*Cyprinus carpio*	Bromex-50	0.35 ppm	I	1	....	Effective	....	Sarig, 1966
,, ,,	*Cyprinus carpio*	Dipterex	0.4 ppm	I	1	....	Effective	....	Sarig, 1966
Dactylogyrus vastator	*Cyprinus carpio*	Bromex-50	0.35 ppm	I	1	....	Effective	....	Lahav, *et al.*, 1966
,, ,,	*Cyprinus carpio*	Dipterex	0.4 ppm	I	1	....	Effective	....	Sarig, 1966; Sarig, *et al.*, 1965
Dactylogyrus sp.	*Cyprinus carpio*	Acriflavine + Ammonium hydroxide (0.25 g Acr./L. of 10% NH_4OH)	100 ppm	D 60–90 sec.	1	....	Effective	....	Ergens, 1962
,, ,,	*Cyprinus carpio*	Ammonia + copper sulfate	25 ppm + 0.5 ppm	I	1	....	Effective	....	Avdosev and Voznyi, 1963
,, ,,	*Cyprinus carpio*	Ammonium hydroxide	2,000 ppm	D 10 min.	1	....	Effective	....	Bauer, 1958

[continued

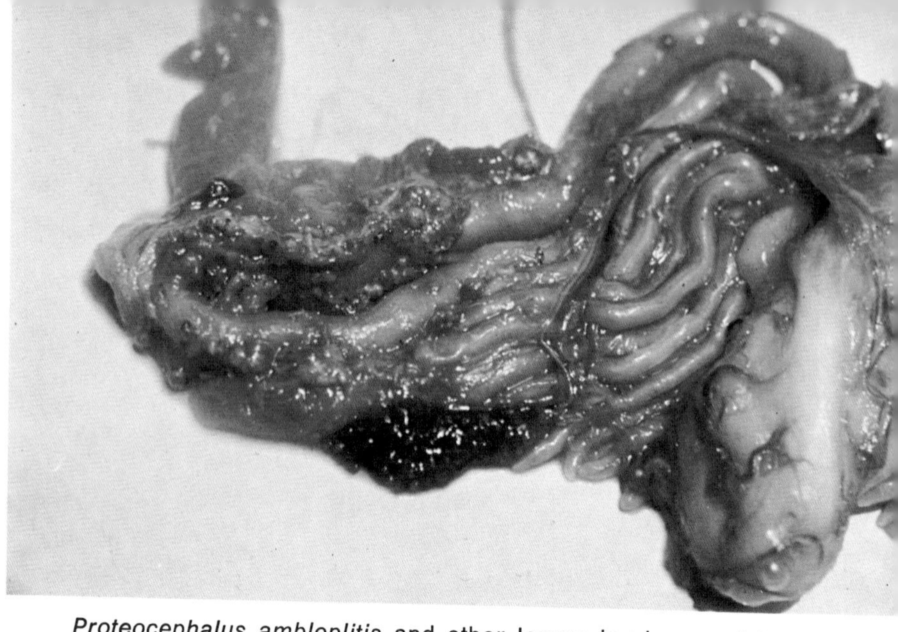

Proteocephalus ambloplitis and other larvae in viscera of largemouth bass. Adhesions in gonads reduce fecundity. Courtesy of Dr. W. Rogers, Auburn University.

Proteocephalus ambloplitis, scolex of plerocercoid. Note 4 suckers and apical organ. Courtesy of Dr. W. Rogers, Auburn University.

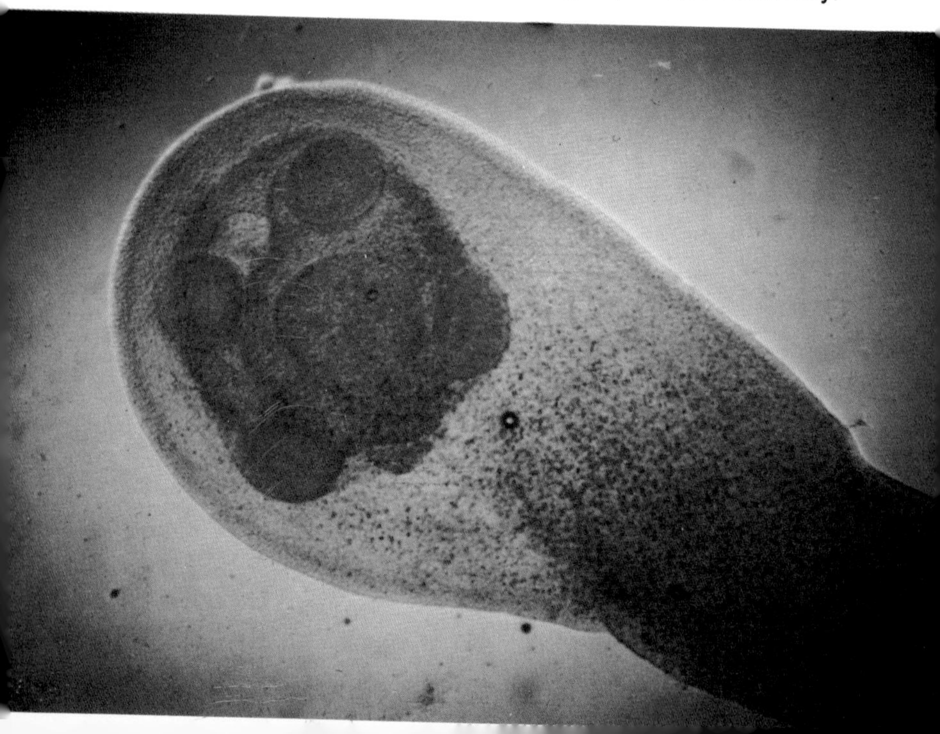

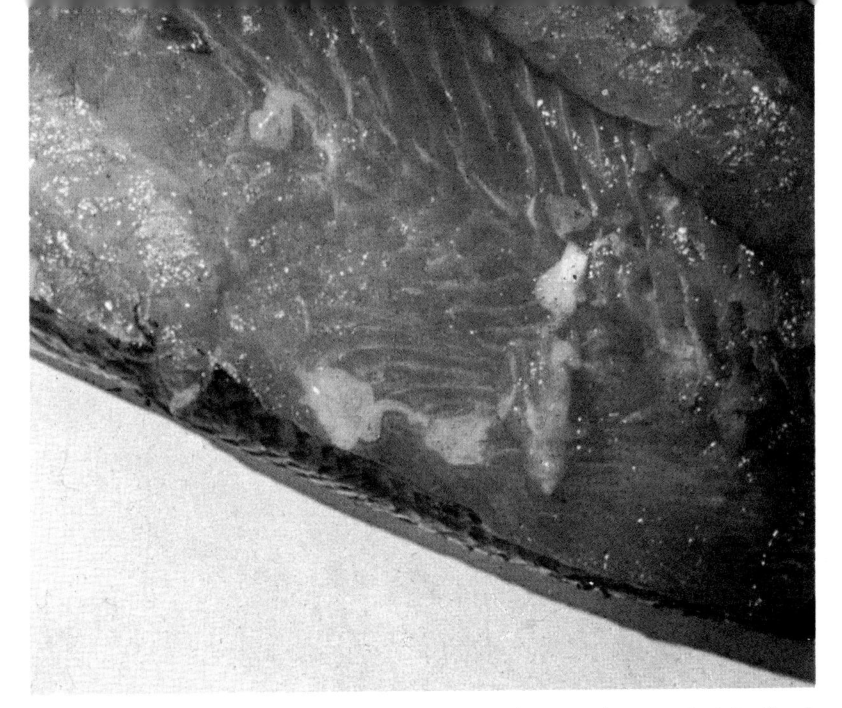

Triaenophorus crassus, plerocercoids free and encysted in flesh *Coregonus lavaretus*. Courtesy of Drs. H. Reichenbach-Klinke and E. Elkan.

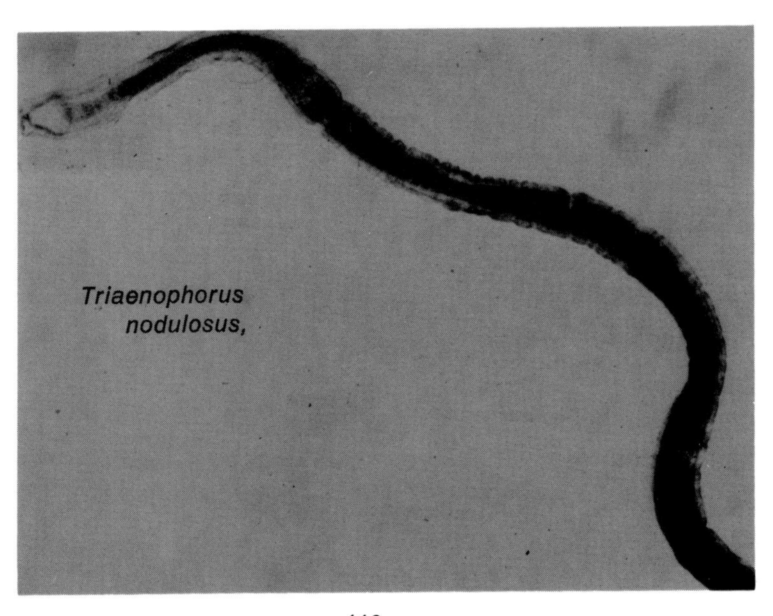

Triaenophorus nodulosus,

Table 6. Monogenetic Trematodes—*continued*

Parasite	Host	Treatment	Dosage	Method	Number of Applications	Frequency	Author's Report of Success	Remarks	References
Dactylogyrus sp.	*Cyprinus carpio*	Ammonium hydroxide	2,000 ppm	D 1 min.	1	....	Effective	Solution only effective for 10 minutes when in use	Lavroskii and Uspenskaya, 1959; Pasovskii, 1953
,, ,,	*Cyprinus carpio*	Ammonium hydroxide	1,000 ppm	D 2 min.	1	....	Effective	....	Ivasik, *et al.*, 1967
,, ,,	*Cyprinus carpio*	Bromex-50	0.12 ppm	I	1	....	Effective	....	Lahav, *et al.*, 1966; Sarig, 1968
,, ,,	*Cyprinus carpio*	DDVP	0.4 ppm	I	1	....	Effective	....	Sarig, 1968
,, ,,	*Cyprinus carpio*	Dipterex (D-50)	0.4 ppm	I	1	....	Effective	....	Sarig, *et al.*, 1965
,, ,,	Unspecified	Drying	....	P	1	....	Effective	Destroys eggs	Hoffman, Unpubl.
,, ,,	*Carassius auratus*	Dylox	0.0625 ppm	I	1	....	Effective	Below 20°C	Osborn, 1966
,, ,,	*Carassius auratus, Notemigonus crysoleucas, Pimephales promelas*	Dylox	0.25 ppm	I	2	3rd day	Effective	Above 33°C	Osborn, 1966; F. Meyer, 1970
,, ,,	*Cyprinus carpio*	Dylox	10,000 ppm	D 15 min.	1	....	Effective	....	Grabda and Grabda, 1968
,, ,,	Salmonids	Formalin	166 ppm	F 1 hr.	1	....	Effective	Below 16°C	Davis, 1953
,, ,,	Salmonids	Formalin	250 ppm	F 1hr.	1	....	Effective	Above 16°C, toxic to weak fish	Davis, 1953
,, ,,	*Carassius auratus*	Formalin	30 ppm	I	2	3rd day	Effective	Recurrences common	Osborn, 1966
,, ,,	*Tinca tinca*	Foschlor	0.1 ppm	I	1	....	Effective	....	Prost and Studnicka, 1968

,,	,,	*Cyprinus carpio*	Magnesium sulfate + sodium chloride	300,000 ppm + 70,000 ppm	D 5–10 min.	1	....	Effective	....	Bauer, 1958
,,	,,	*Cyprinus carpio*	Masoten	0.25 ppm	I	1	....	Effective	In ponds	Plate, 1970
,,	,,	*Cyprinus carpio*	Masoten	25,000 ppm	D 5–10 min.	1	....	Effective	....	Plate, 1970
,,	,,	*Carassius auratus*	Neguvon	1.0 ppm	F 48 hrs.	1	....	Effective	....	Bailosoff, 1963
,,	,,	*Cyprinus carpio*	Neguvon	25,000 ppm	D 3 min.	1	....	Effective	....	Bailosoff, 1963
,,	,,	*Cyprinus carpio*	Oxygen depletion	....	P	1	....	Effective	....	Bauer, 1959
,,	,,	*Carassius auratus*	Potassium permanganate	5 ppm	I	1	....	Effective	....	Hess, 1930
,,	,,	*Carassius auratus*	Potassium permanganate	3.8 ppm	F 2 hrs.	1	....	Effective	....	Hess, 1930
,,	,,	*Carassius auratus*	Potassium permanganate	460 ppm	D 1½–2 min.	1	....	Effective	....	Hess, 1930
,,	,,	Trout, *Carassius auratus*	Potassium permanganate	10 ppm	F 1 hr.	1	....	Effective	Continuous flow for one hour	Kingsbury and Embody, 1932
,,	,,	Trout	Potassium permanganate	3.3 ppm	D 30 min.	1	....	Effective	....	Prevost, 1934
,,	,,	*Carassius auratus*	Potassium permanganate	10 ppm near shore	I	1	....	Effective	Spray 10 ft. from shoreline	Meehean, 1937
,,	,,	*Notemigonus crysoleucas, Pimephales promelas*	Potassium permanganate	2 ppm	I	1	....	Not effective	....	F. Meyer, Unpubl.

[continued

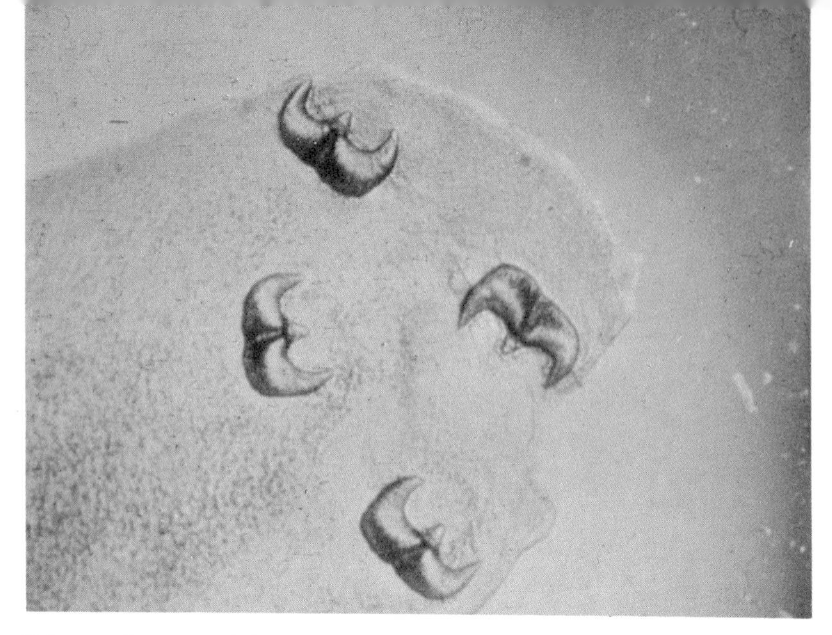

Triaenophorus nodulosus, scolex of plerocercoid showing tricuspid hooks.

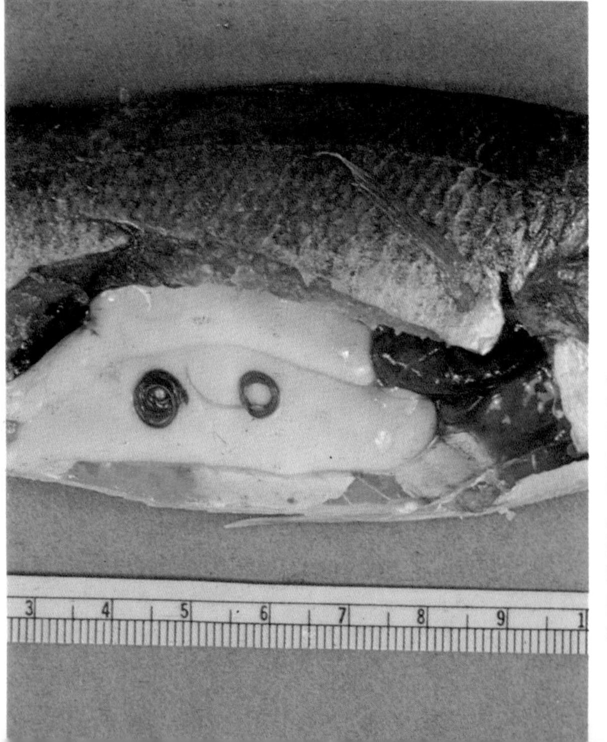

Eustrongylides (red nematode larvae) in yellow perch. Courtesy of Dr. A. Dechtiarenko, Fisheries Research Lab, Maple, Ontario, Canada.

Philometra nodulosa in "cheek galleries" of *Catostomus commersoni*. Collected by J. Brown, Martinsburg, West Virginia. Photo by G. Hoffman.

Philometra sp. in eye of bluegill. Courtesy of Dr. T. Wellborn, Mississippi State University.

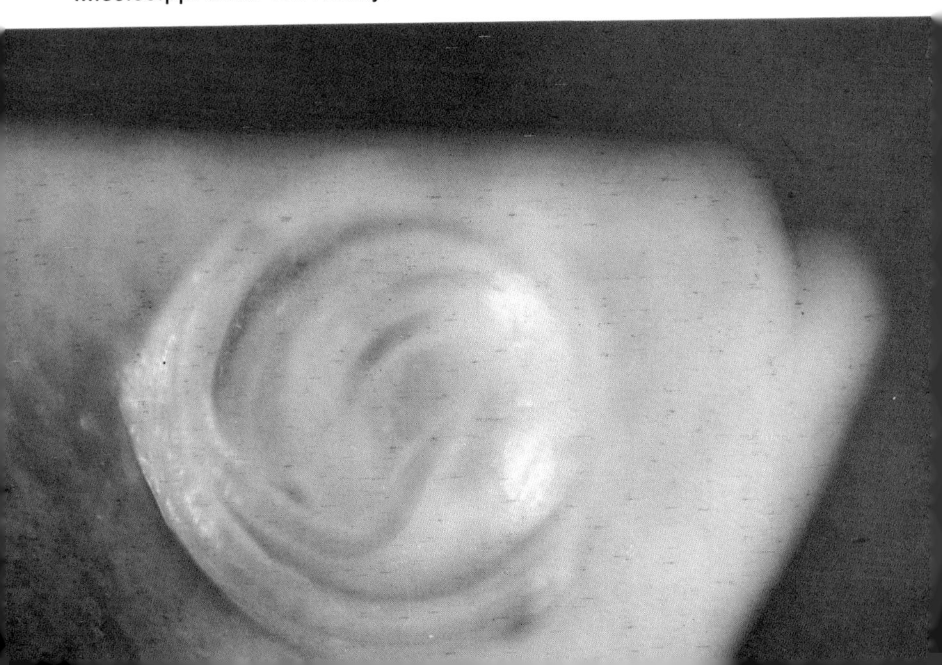

Table 6. Monogenetic Trematodes—*continued*

Parasite	Host	Treatment	Dosage	Method	Number of Applications	Frequency	Author's Report of Success	Remarks	References
Dactylogyrus sp.	*Cyprinus carpio*	Quicklime	2,000 ppm	D 5 sec.	1	....	Effective	Detrimental effect on fish	Schäperclaus, 1954
,, ,,	Unspecified	Sodium chlorate	3,000 ppm	?	1	....	Effective	Separate fry from adults	Popov and Jankov, 1968
,, ,,	Unidentified	Sodium chloride	25,000 ppm	F 20 min.	As needed	Daily	Effective	Avoid zinc containers	Amlacher, 1961b
,, ,,	*Salmo trutta*	Sonic vibrations	380 khz	P 25 sec.	1	....	Effective	Used on fry	Nechaeva, 1959
,, ,,	*Cyprinus carpio*	Ultraviolet light	MBU-3 (PRK-7)	P	....	Continuous	Effective	Dosage not given	Kokhanskaya, 1970
Discocotyle sp.	Salmonids	Zonite	265 ppm	D 2 min.	1	....	Effective	....	Laird and Embody, 1931
Gyrodactylus elegans	*Cyprinus carpio*	Quinine hydrochloride	20 ppm	F 24 hrs.	1	....	Not effective	Toxic to fish	Schäperclaus, 1954
Gyrodactylus macrochiri	*Lepomis macrochirus*	Heat	Raise to 21°C	P	1	....	Effective	Raise water temp. to 21°C	Hoffman and Putz, 1964
Gyrodactylus wegneri	*Notemigonus crysoleucas*	Dylox	0.25 ppm	I	1	....	Effective	....	F. Meyer, 1970
,, ,,	*Notemigonus crysoleucas*	Formalin	50 ppm	F 14 hrs.	1	....	Effective	....	Lewis & Lewis, 1963
,, ,,	*Notemigonus crysoleucas*	Paraformaldehyde	10 ppm	I	1	....	Effective	Dissolve in soda ash	Lewis and Parker, 1965; S. Lewis, 1967
Gyrodactylus sp.	Trout	Acetic acid	2,000 ppm	D 1 min.	1	....	Effective	Increase time if large numbers of fish used	Embody, 1924

,,	,,	Aquarium fish, *Cyprinus carpio*	Ammonium chloride	25,000 ppm	D 10–15 min.	?	Daily	Effective	Repeat as needed	Roth, 1928; Reichenbach-Klinke, 1966; Schäperclaus, 1954; Van Duijn, 1967
,,	,,	*Cyprinus carpio*	Bromex-50	0.12–0.18 ppm	I	1	....	Effective	....	Lahav, *et al.*, 1966; Sarig, 1968
,,	,,	*Cyprinus carpio*	Chloramine	10 ppm	F 24 hrs.	1	....	Effective	....	Schäperclaus, 1954
,,	,,	Unspecified	Chloramine-B	10 ppm	F 25 hrs.	1	....	Effective	....	Goncharov, 1966
,,	,,	*Notemigonus crysoleucas*	Chlorine	2.5 ppm	D 10 sec.	1	....	Effective	....	Lewis and Ulrich, 1967
,,	,,	Unspecified	Copper sulfate	100 ppm	F 30 min.	1	....	Effective	....	Schäperclaus, 1954; Amlacher, 1961b
,,	,,	Pond fishes	Copper sulfate	1 ppm	I	1	....	Effective	Toxic in soft waters	Wellborn, 1967
,,	,,	*Cyprinus carpio*	Dipterex (D-50)	0.4 ppm	I	1	....	Effective	....	Sarig, *et al.*, 1965
,,	,,	*Carassius auratus*	Dylox	0.06 ppm	I	1	....	Effective	Below 20°C	Osborn, 1966
,,	,,	*Carassius auratus*	Dylox	0.25 ppm	I	1	....	Effective	Above 33°C	Osborn, 1966
,,	,,	*Notemigonus crysoleucas*, Pond fishes	Formalin	25 ppm	I	1	....	Effective	4° to 16°C	Lewis and Lewis, 1963; Wellborn, 1967
,,	,,	*Notemigonus crysoleucas*	Formalin	50 ppm	F 14 hrs.	1	....	Effective	4° to 16°C	Lewis and Lewis, 1963
,,	,,	*Esox* sp., *Salmo* sp.	Formalin	250 ppm	F 30 min.	?	Alternate days	Effective	Treat as needed	O'Brien (in Mellen, 1928)

[*continued*

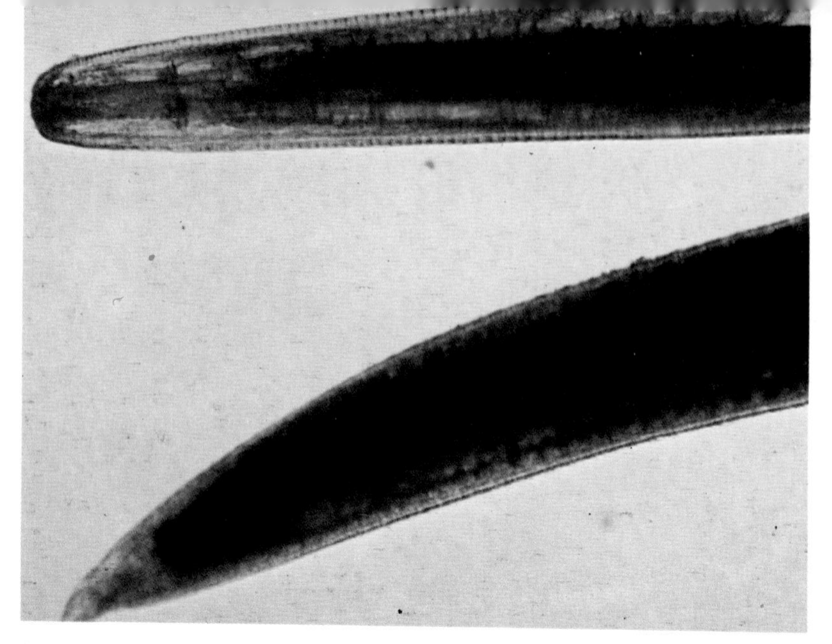

Philometra abdominalis, red worm from body cavity of European cyprinids. Photo by Frickhinger.

Philometra sp. giving birth to larvae which must be eaten by copepods to continue the life cycle. Photo by Frickhinger.

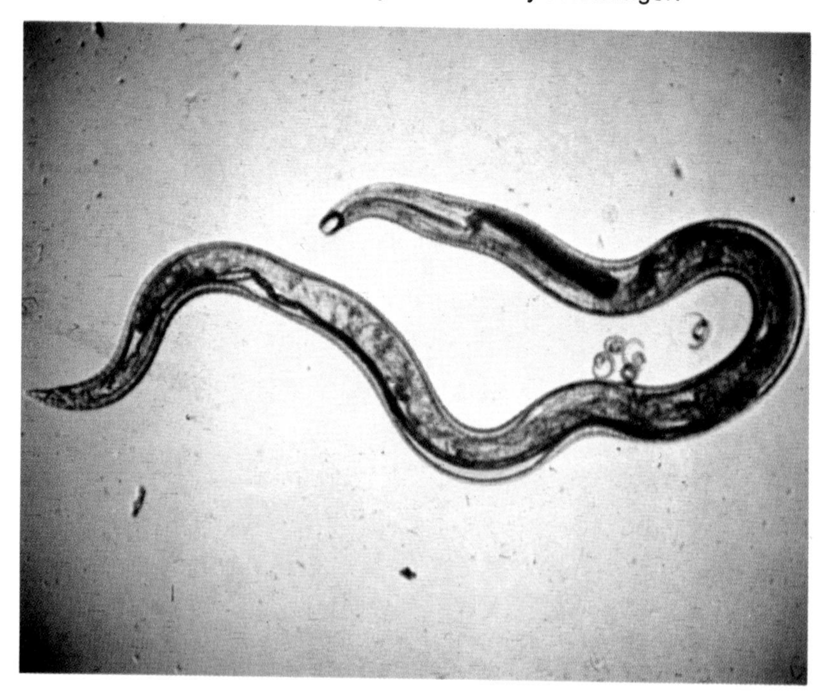

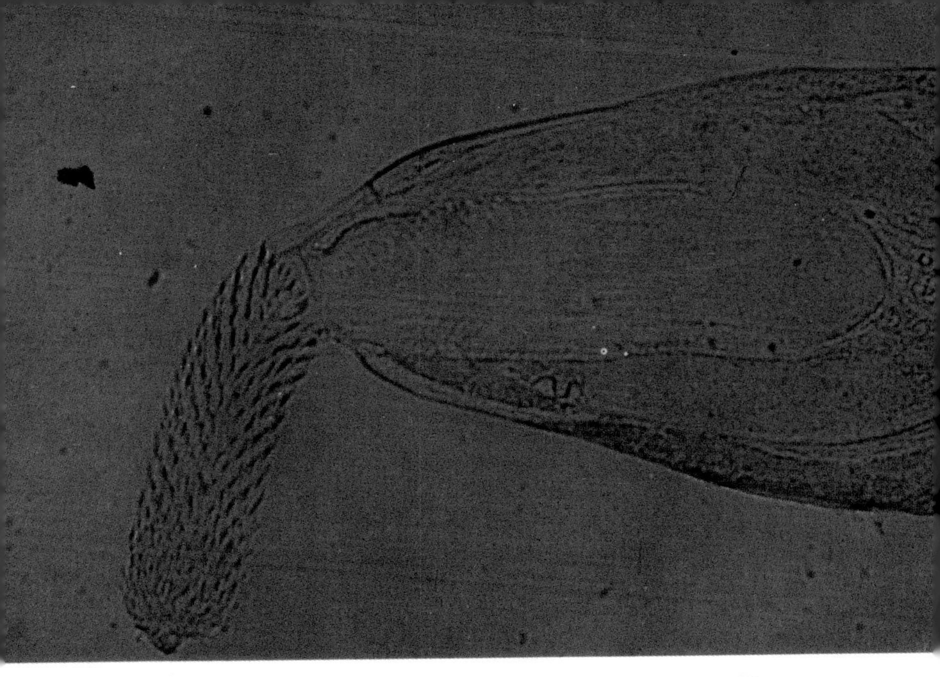

Echinorhynchus truttae. Adult acanthocephalan from trout. Photo by G. Hoffman.

Leptorhynchoides thecatus. Adult acanthocephalan from intestine of black bullhead. Photo by F. Meyer.

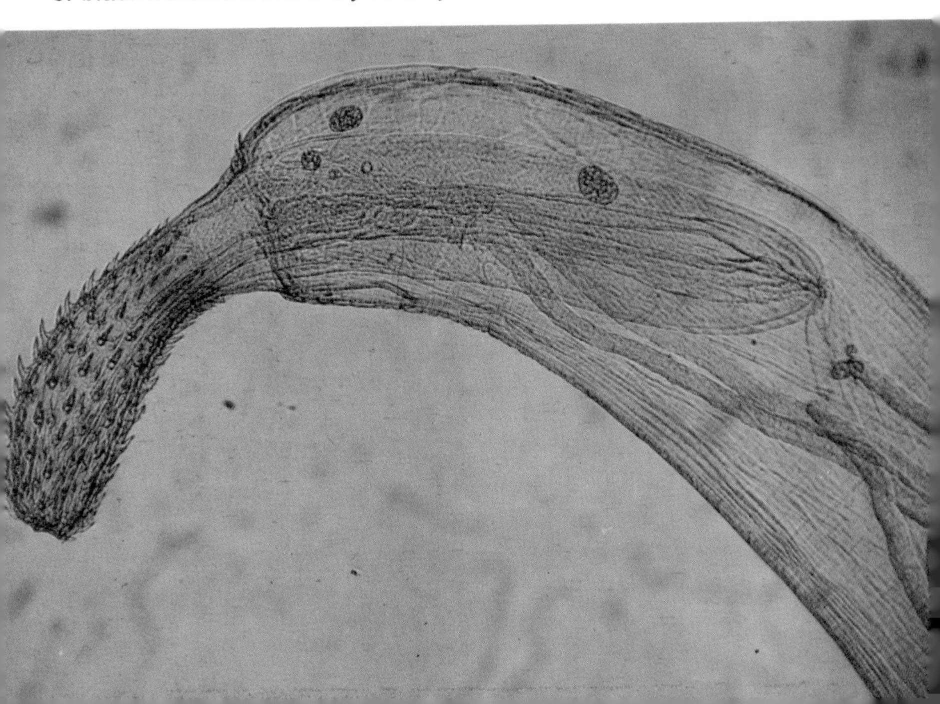

Table 6. Monogenetic Trematodes—*continued*

Parasite	Host	Treatment	Dosage	Method	Number of Applications	Frequency	Author's Report of Success	Remarks	References
Gyrodactylus sp.	Trout	Formalin	166 ppm	F 1 hr.	1	....	Effective	Repeat as needed	Fish, 1940
,, ,,	*Ictalurus nebulosus*	Formalin	10–15 ppm	I	1	....	Effective	....	R. Allison, 1957a and 1963
,, ,,	*Carassius auratus*	Formalin	15 ppm	I	1	....	Effective	....	Kumar, 1958
,, ,,	Trout	Formalin	250 ppm	F 1 hr.	1	....	Effective	....	Davis, 1953
,, ,,	Unspecified	Formalin	200 ppm	F 15 min.	1	....	Effective	....	Schäperclaus, 1954
,, ,,	*Fundulus heteroclitus*	Fresh water	100%	F 5 hrs.	1	....	Effective	Marine parasite	Gowanloch, 1927
,, ,,	*Cyprinus carpio*	Lysol	200 ppm	D 30 sec.	1	....	Effective	Not as good as Formalin	Schäperclaus, 1954
,, ,,	*Cyprinus carpio*	Malachite green	0.16 ppm	I	1	....	Effective	Winter ponds	Sokolov and Maslyukova, 1971
,, ,,	*Cyprinus carpio*	Masoten	0.25–4.0 ppm	I	1	....	Effective	....	Plate, 1970
,, ,,	Unspecified	Mercurochrome	10 ppm	F 12 hrs.	1	....	Effective	....	Seale, 1928
,, ,,	*Cyprinus carpio*	Methylene blue	3 ppm	I	?	....	Not effective	....	Schäperclaus, 1954
,, ,,	Unspecified	Methylene blue	3 ppm	I	3	Daily	Effective	....	Reichenbach-Klinke, 1966
,, ,,	Unspecified	Methylene blue	5 ppm	I	1	....	Effective	....	Van Duijn, 1956
,, ,,	Unspecified	Methylene blue	3 ppm	I	1	....	Not effective	....	Schäperclaus, 1954; Amlacher, 1961
,, ,,	*Cyprinus carpio*	Neguvon	25,000 ppm	D 3 min.	1	....	Effective	....	Bailosoff, 1963
,, ,,	*Notemigonus crysoleucas*	Paraformaldehyde	10 ppm	I	1	....	Effective	Dissolve in soda ash	Lewis and Parker, 1965

,,	,,	*Micropterus salmoides*	*Pimephales promelas*	Not given	B	1	....	Not given	Minnow eats *Gyrodactylus*	Spall, 1970
,,	,,	*Salvelinus fontinalis*	Potassium permanganate	13.5 ppm	F 1 hr.	1	....	Effective	May need to be repeated	L. Wolf, 1935a and b
,,	,,	*Carassius auratus*	Potassium permanganate	10 ppm	I	1	....	Effective	Treat only outer margin of pond	Meehean, 1937
,,	,,	*Carassius auratus*	Potassium permanganate	3–5 ppm	I	1	....	Effective	....	Hess, 1930; Kumar, 1958
,,	,,	*Notemigonus crysoleucas*	Potassium permanganate	2 ppm	I	1	....	Partially effective	May need to be re-treated	F. Meyer, Unpubl.
,,	,,	Unspecified	Potassium permanganate	10 ppm	I	?	....	Effective	....	Hofer, 1928
,,	,,	*Carassius auratus*	Potassium permanganate	3.8 ppm	F 2 hrs.	1	....	Effective	....	Hess, 1930
,,	,,	*Carassius auratus*	Potassium permanganate	460 ppm	D $1\frac{1}{2}$–2 min.	1	....	Effective	....	Hess, 1930
,,	,,	Trout	Potassium permanganate	3.3 ppm	F 30 min.	1	....	Effective	....	Prevost, 1934
,,	,,	Trout	Potassium permanganate	10 ppm	F 1 hr.	1	....	Effective	Constant flow	Fish, 1933
,,	,,	*Cyprinus carpio*	Quicklime	2,000 ppm	D 5 sec.	1	....	Effective	Detrimental effect on fish	Schäperclaus, 1954
,,	,,	*Cyprinus carpio, Lebistes reticulatus,* Unspecified	Roccal (Zephiran)	250–500 ppm	F 20–30 min.	1	....	Effective	....	Schäperclaus, 1954; Reichenbach-Klinke, 1966
,,	,,	*Cyprinus carpio*	Salicylic acid	11 ppm	F 30 min.	1	....	Effective	....	Hofer, 1928

[*continued*

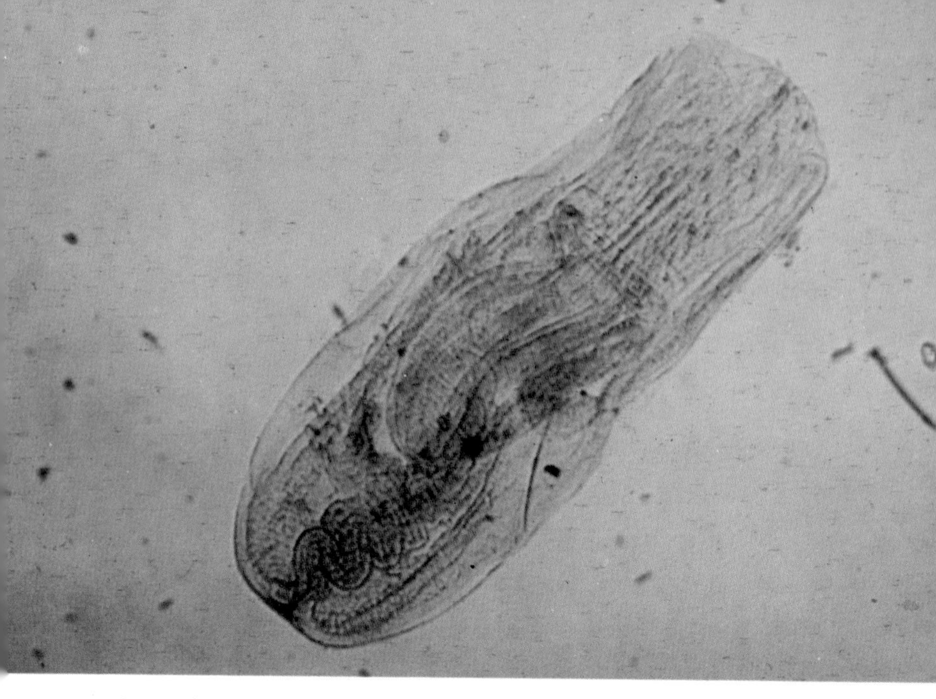

Leptorhynchoides thecatus. Cystocanth from mesenteries of black bullhead. Photo by F. Meyer.

Octospiniferoides chandleri, probscis hooks at high magnification. Courtesy of Dr. W. Bullock, University of New Hampshire.

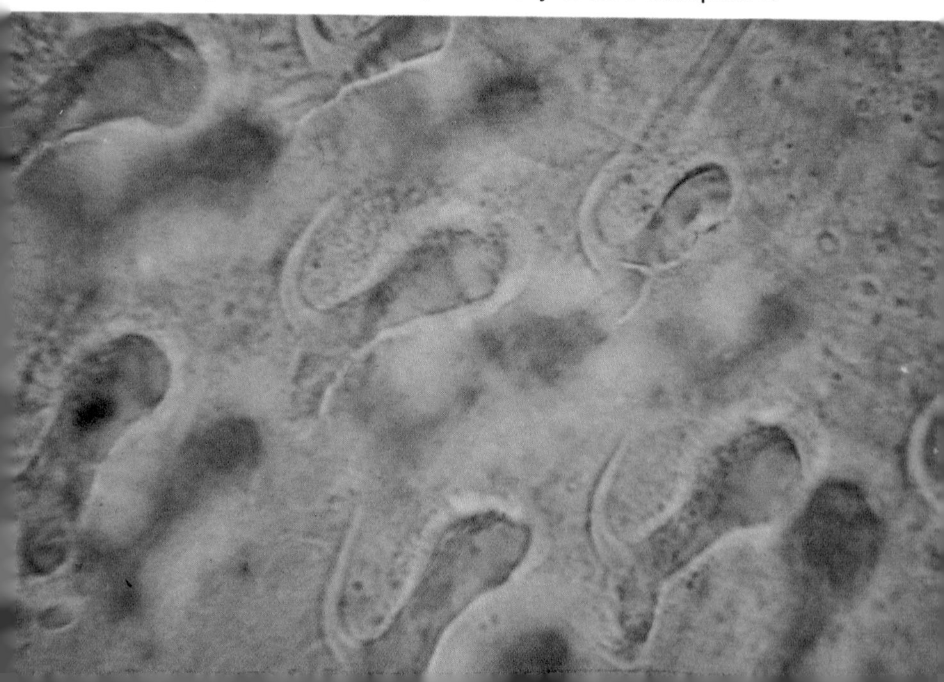

Pomphorhynchus bulbocolli, adult acanthocephala in intestine of striped bass. Courtesy of Dr. T. Wellborn, Mississippi State University.

Pomphorynchus bulbocolli. Tissue section showing how the proboscis (P) and "bulb" become inserted deep in the tissues of the intestine. Courtesy of Dr. W. Bullock, University of New Hampshire.

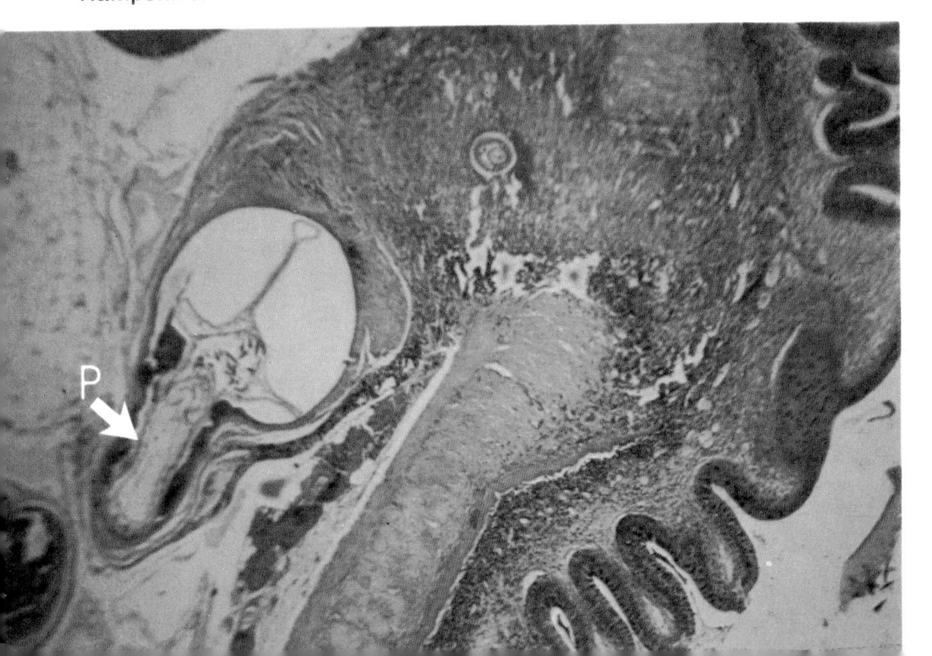

Table 6. Monogenetic Trematodes—*continued*

Parasite	Host	Treatment	Dosage	Method	Number of Applications	Frequency	Author's Report of Success	Remarks	References
Gyrodactylus sp.	Trout	Sodium chloride	50,000 ppm	D $1\frac{1}{2}$–$2\frac{1}{2}$ min.	1	....	Effective	....	Guberlet, *et al.*, 1927
,, ,,	Aquarium fish	Sodium chloride	30,000 ppm	F 15–30 min.	1	....	Effective	....	Van Duijn, 1967
,, ,,	Aquarium fish	Sodium chloride	17,000 ppm	I	4	Daily	Effective	Gradually increase concentration over a 4 day period	Van Duijn, 1967
,, ,,	*Salmo trutta*	Sonic vibrations	380 khz	P 25 sec.	1	....	Effective	Used on fry	Nechaeva, 1959
Microcotyle sp.	Unspecified	Silvol	20,000 ppm	D 3 min.	1	....	Effective	....	Mellen, 1928
Unidentified Monogenea	Unspecified	Acetic acid	20,000 ppm	D 1 min.	1	....	Effective	....	Embody, 1928; Davis, 1953
,,	*Cyprinus carpio*	Ammonium chloride	25,000 ppm	D 10–15 min.	1	....	Effective	....	Roth, 1928; Reichenbach-Klinke, 1966
,, ,,	*Cyprinus carpio*	Ammonium hydroxide plus copper sulfate	25 ppm + 0.4–0.5 ppm	I	1	....	Effective	Treat feeding area	Advosev & Voznyi, 1963
,, ,,	*Cyprinus carpio*	Ammonium hydroxide	2,000 ppm	D 10 min.	1	....	Effective	....	Pasovskii, 1953; Lavroskii and Uspenskaya, 1959; Reichenbach-Klinke, 1966
,, ,,	*Cyprinus carpio*	Ammonium hydroxide	500 ppm	D 5–15 min.	1	....	Effective	Do not use on weak fish	Van Duijn, 1956
,, ,,	*Cyprinus carpio*	Chloramine-B	10 ppm	F 25 hrs.	1	....	Effective	....	Van Duijn, 1967

,,	,,	*Cyprinus carpio*	Copper sulfate	1 ppm	I	1	....	Effective	Use only if water hardness is over 100 ppm as $CaCO_3$	Wellborn, 1967
,,	,,	*Cyprinus carpio*	Copper sulfate	100 ppm	F10–30 min.	1	....	Effective	....	Amlacher, 1961b
,,	,,	Unspecified	Drying	....	P-M	1	....	Effective	....	Hoffman, Unpubl.
,,	,,	Unspecified	Dylox	0.4 ppm	F 48 hrs.	1	....	Effective	....	Bailosoff, 1963
,,	,,	Unspecified	Dylox	1.0 ppm	I	1	....	Effective	Kills adults but not eggs	Bailosoff, 1963
,,	,,	Unspecified	Dylox	0.4 ppm	I	1	....	Effective	....	Wellborn, 1967
,,	,,	Unspecified	Formalin	15–25 ppm	I	1	....	Effective	....	Fish and Burrows, 1940; Allison, 1957a; Kumar, 1958
,,	,,	*Notemigonus crysoleucas*, Salmonids	Formalin	25 ppm	I	1	....	Effective	....	Snow, 1962; Lewis and Lewis, 1963; Rucker, *et al.*, 1963; F. Meyer, 1966b; Wellborn, 1967
,,	,,	Unspecified	Hydrogen peroxide	5,000 ppm	D 5 min.	1	....	Effective	....	Plehn (in Schäperclaus, 1954)
,,	,,	Unspecified	Hydrogen peroxide	30,000 ppm	D 10 min.	1	....	Effective	....	Roth (in Schäper-claus, 1954)
,,	,,	Aquarium fishes	Hydrogen peroxide	30 ppm	I	1	....	Effective	....	Dempster and Shipman, 1970

[continued

Leeches attacking channel catfish fingerlings. Photo by G. Hoffman.

Leeches attacking Great Lakes smelt. Courtesy of Dr. A. Dechtiarenko, Fisheries Research Lab, Maple, Ontario, Canada.

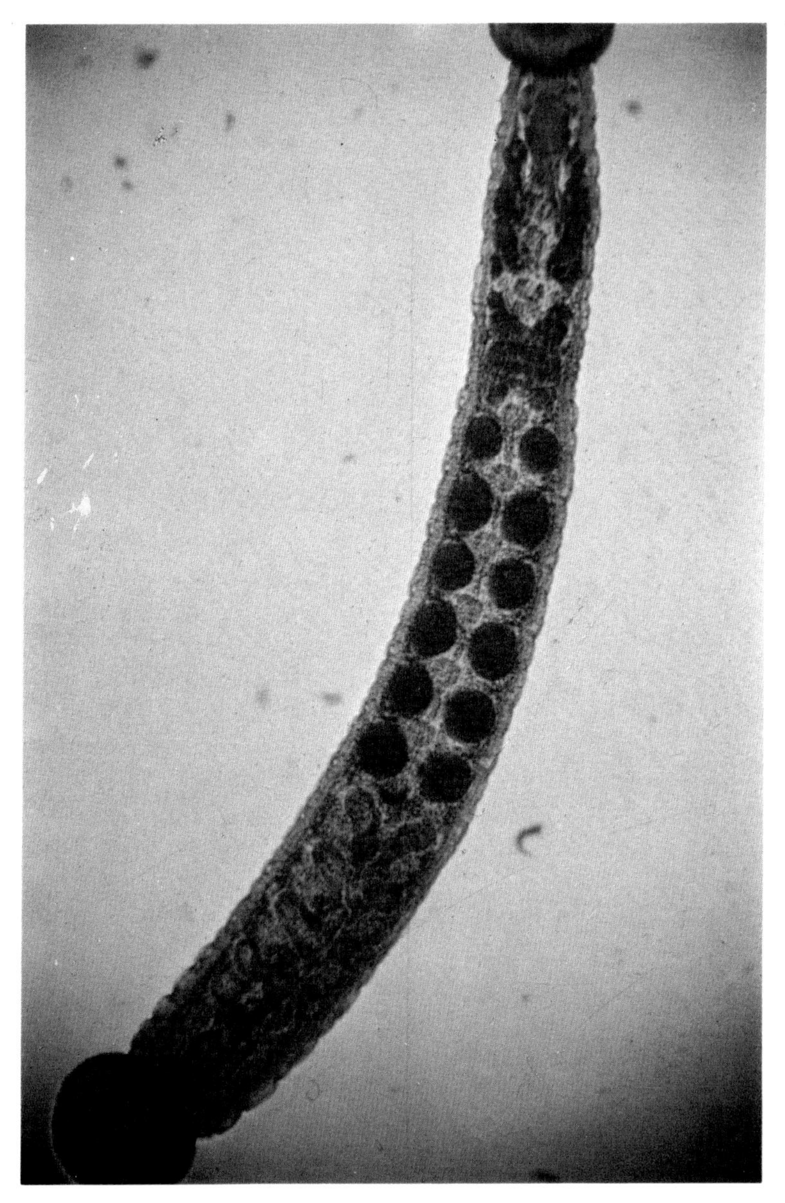

Piscicola sp., leech from skin of fish. Note large suckers and 6 pairs of spherical testes. Stained specimen.

Table 6. Monogenetic Trematodes—*continued*

Parasite	Host	Treatment	Dosage	Method	Number of Applications	Frequency	Author's Report of Success	Remarks	References
Unidentified Monogenea	Aquarium fishes	Magnesium sulfate + sodium chloride	30,000 ppm $MgSO_4$ + 7,000 ppm NaCl	F 5–10 min.	1	....	Effective	....	Bauer, 1958
,, ,,	Aquarium fishes	Mercurochrome	1–2 ppm	I	1	....	Effective	....	Detwiler and McKennon, 1929
,, ,,	Unspecified	Mercurochrome	10 ppm	F 12 hrs.	1	....	Effective	....	Seale, 1928
,, ,,	Unspecified	Methylene blue	2–4 ppm	I	1	....	Effective	....	Amlacher. 1961b
,, ,,	*Carassius auratus*	Methylene blue	25 ppm	I	1	....	Effective	....	Kumar, 1958
,, ,,	Unspecified	Methylene blue	2–4 ppm	I	1	....	Effective	....	Van Duijn, 1967
,, ,,	Unspecified	Methylene blue	3 ppm	I	1	....	Effective	....	Reichenbach-Klinke, 1966
,, ,,	Unspecified	Picric acid	20 ppm	F 1 hr.	1	....	Effective	....	Reichenbach-Klinke, 1966; Van Duijn, 1967
,, ,,	Unspecified	Potassium antimony tartrate	1.5 ppm	I	1	....	Effective	In aquariums	Reichenbach-Klinke, 1966; Van Duijn, 1967
,, ,,	Unspecified	Potassium permanganate	5 ppm	I	1	....	Effective	....	Reichenbach-Klinke, 1966
,, ,,	Salmonids, unspecified	Potassium permanganate	5–10 ppm	F 1–2 hrs.	1	....	Effective	....	Hess, 1930; Kingsbury and Embody, 1932; Fish, 1933

Parasites	Host	Treatment	Dosage	Method	Number of Applications	Frequency	Author's Report of Success	Remarks	References
,, ,,	Salmonids	Potassium permanganate	3–5 ppm	I	1	....	Effective	....	Kumar, 1958; Hess, 1930
,, ,,	Trout	PMA (pyridylmercuric acetate)	2 ppm	F 1 hr.	1	....	Effective	Toxic to *Salmo gairdneri*, hazardous to human; banned by USDI	Reichenbach-Klinke, 1966
,, ,,	Unspecified	Rivanol	2–2.5 ppm	D	1	....	Effective	....	Amlacher, 1961b; Reichenbach-Klinke, 1966
,, ,,	Unspecified	Sodium chloride	50,000 ppm	D 90 sec.	1	....	Effective	....	Guberlet, *et al.*, 1927
,, ,,	Unspecified	Sodium chloride	30,000 ppm	F 15–30 min.	1	....	Effective	....	Van Duijn, 1967

Table 7. DIGENETIC TREMATODES

Parasites	Host	Treatment	Dosage	Method	Number of Applications	Frequency	Author's Report of Success	Remarks	References
Alloglossidium sp.	*Ictalurus punctatus*	Di-N-butyl tin oxide	250 mg/kg of fish	O	3	Daily	Effective	....	R. Allison, 1957b
Crepidostomum farionis	*Salmo aquabonita*	Di-N-butyl tin oxide	250 mg/kg of fish	O	3	Daily	Effective	....	Mitchum and Moore, 1966
Nanophyetus salmincola cercariae	Salmonids	Electrical grid	....	P	....	Daily	Effective	Kills cercariae in incoming water	Combs, 1968
Cercariae	Salmonids	Sand-gravel filters	....	P	....	Daily	Effective	....	Burrows and Combs, 1968

Argulus giordanii on European eel. Courtesy of Dr. P. Ghittino, Istituto Zooprofilattico Sperimentale del Piemonte e della Liguria, Torino, Italy, and Edizioni Rivista di Zootecnia.

Argulus sp. on three-spined stickleback.

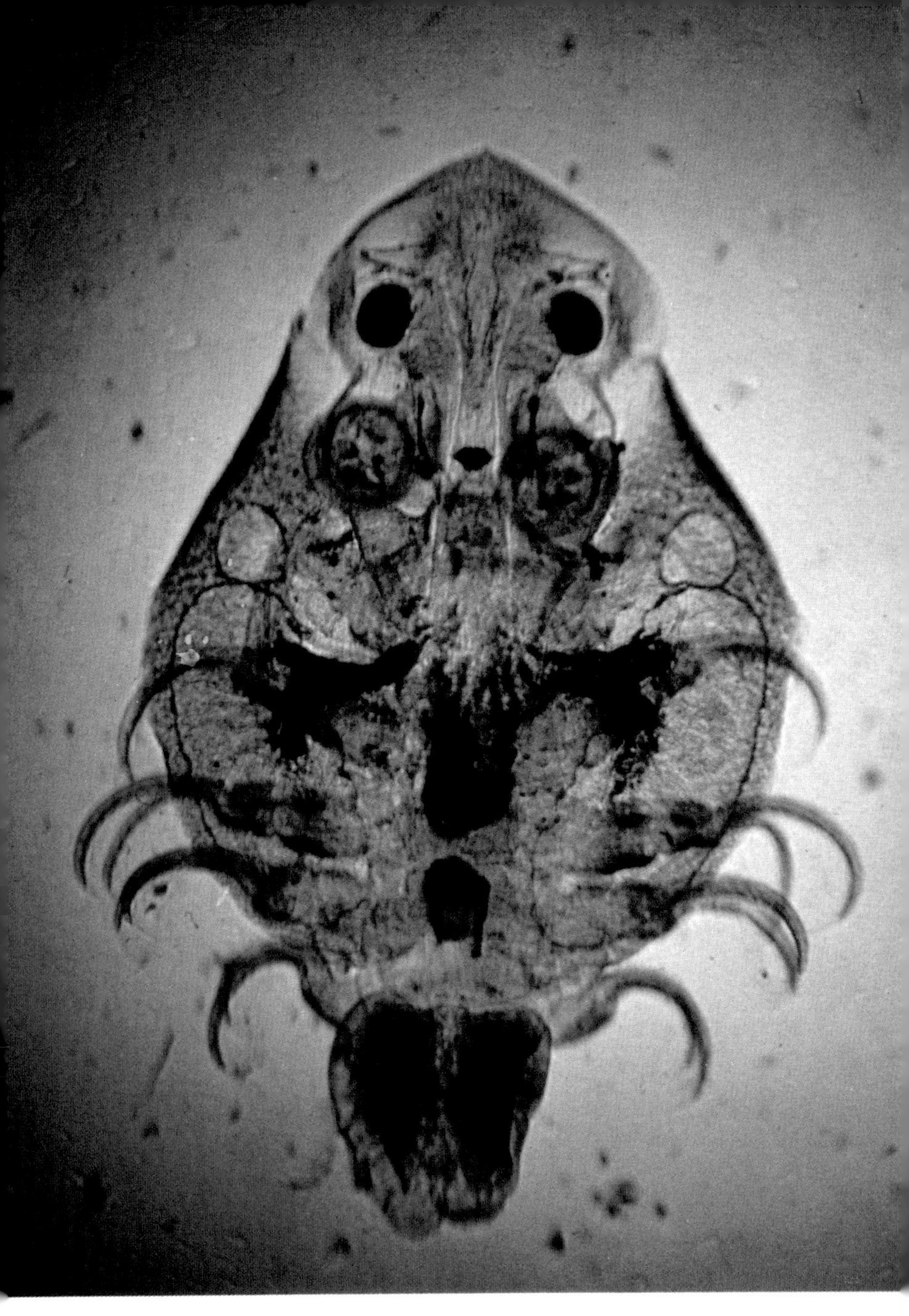

Argulus sp. Ventral view, magnified about 50 times. Photo by Dr. E. Elkan.

Table 8. CESTODES

Parasite	Host	Treatment	Dosage	Method	Number of Applications	Frequency	Author's Report of Success	Remarks	References
Bothriocephalus gowkongensis	*Cyprinus carpio*	Phenasal	1 g/kg of fish	0	6	Daily	Effective	Feed as pellets	Muzykovski, 1968; 1971
Bothriocephalus sp.	*Ctenopharyngodon idellus*	Bithionol	10,000 ppm in food	0	3	Daily	Not effective	Caused hematological disorders	Klenov, 1970
,, ,,	*Ctenopharyngodon idellus*	Calcium chloride	Not given	M	?	?	Effective	....	Babaev and Shcherbakova, 1963
,, ,,	*Cyprinus carpio*	Fenasal (phenasal)	10,000 ppm in food	0	3	Daily	Effective	Expect re-infection in one month	Nazarova, *et al.*, 1969
,, ,,	*Ctenopharyngodon idellus*	Fenasal (phenasal)	10,000 ppm in food	0	1	....	Effective	....	Klenov, 1970
,, ,,	*Cyprinus carpio*	Kamala	75,000 ppm in food	0	1	....	Not effective	....	Nazarova, *et al.*, 1969
,, ,,	*Cyprinus carpio*	Phenothiazine	75,000 ppm in food	0	1	....	Not effective	....	Nazarova, *et al.*, 1969
,, ,,	*Cyprinus carpio*	Phenothiazine	4–5 g/kg of fish	0	3	Daily	Effective	....	Kulakovskaya and Musselius, 1962
,, ,,	*Cyprinus carpio*	Phenothiazine	4–5 g/kg of fish	0	3	Daily	Effective	....	Kanaev, 1967
,, ,,	*Ctenopharyngodon idellus*	Sodium fluosilicate	10,000 ppm in food	0	3	Daily	Not effective	....	Klenov, 1970
Caryophyllaeus sp.	*Cyprinus carpio*	Kamala	2% of diet	0	7	Daily	Effective	....	Bauer, 1959
,, ,,	*Cyprinus carpio*	Lime pond bottom and banks	Not given	P-M	....	....	Partially effective	Controls *Tubifex* intermediate host	Bauer, 1959

Corallobothrium fimbriatum	*Ictalurus punctatus*	Di-N-butyl tin oxide	250 mg/kg of fish	0	5	Daily	Effective	Repeat in 1 month	F. Meyer, Unpubl.
,, ,,	*Ictalurus punctatus*	Kamala	2% of diet	0	7–14	Daily	Effective	Not as effective as Di-N-butyl tin oxide	R. Allison, 1957b
Diphyllobothrium sp. larvae	Salmonids	Removal of adult host	....	M	....	Daily	Partially effective	Rob eggs from gull nests	Vik, 1965
Eubothrium sp.	*Salmo gairdneri*	Di-N-butyl tin oxide	250 mg/kg of fish	0	3	Daily	Effective	....	Hnath, 1970
,, ,,	*Salmo gairdneri*	Di-N-butyl tin oxide	250 mg/kg of fish	0	3	Daily	Effective	Does not kill worm eggs	Kerr, 1969
Proteocephalus ambloplitis	*Micropterus dolomieui*	Bayer 2353	226 mg/lb of fish	0	1	....	Effective	Used as oral drench	Larsen, 1964
,, ,,	*Micropterus salmoides*	Use parasite-free broodstock	....	M	....	....	Effective	Use parasite-free broodstock	Hunter, 1942; Becker and Brunson, 1968
Proteocephalus sp.	*Salmo gairdneri*	Kamala	1.5–2% of diet	0	10–14	Daily	Effective	Repeat 2 or 3 times at weekly intervals	McKernan, 1940
Triaenophorus sp.	*Coregonus* sp.	Remove adult host	....	M	....	....	Partially effective	Remove *Esox lucius*, host for adult worms	Lawler, 1959
Unidentified Cestodes	Unspecified	Dibutyltin dilaurate	250–500 mg/kg of fish	0	3	Daily	....	Not tried on fish; is an effective, available veterinary product	Hoffman, Unpubl.

[continued

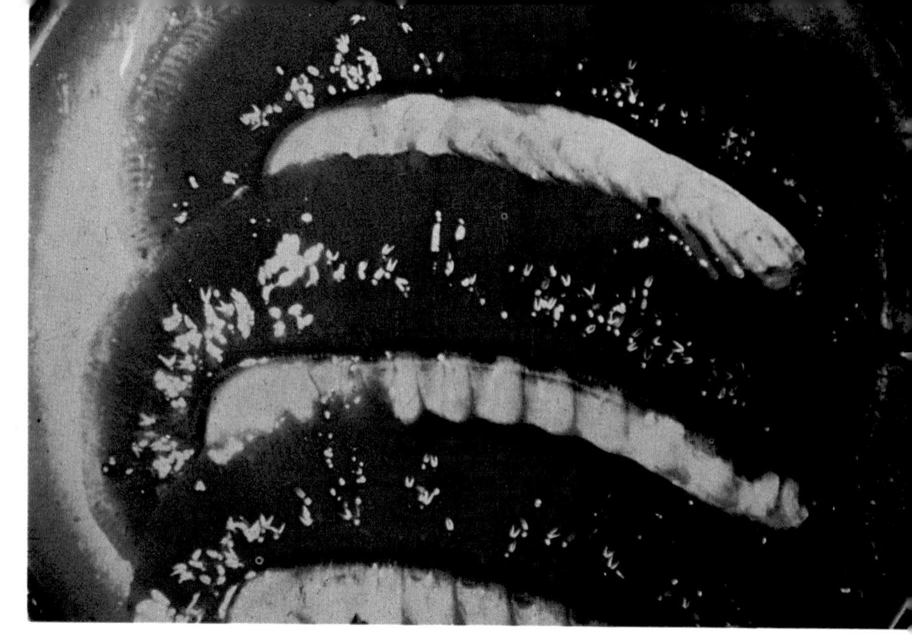

Ergasilus sp. on gills of bream. Parasites are white with two white egg sacs. Courtesy of P. de Kinkelin, Laboratoire d'Ichthyopathologie, Route de Thiverval, 78 Thiverval-Grignon, France.

Ergasilus sp. attached to gill of white crappie. Claspers are embedded in gill. Photo by F. Meyer.

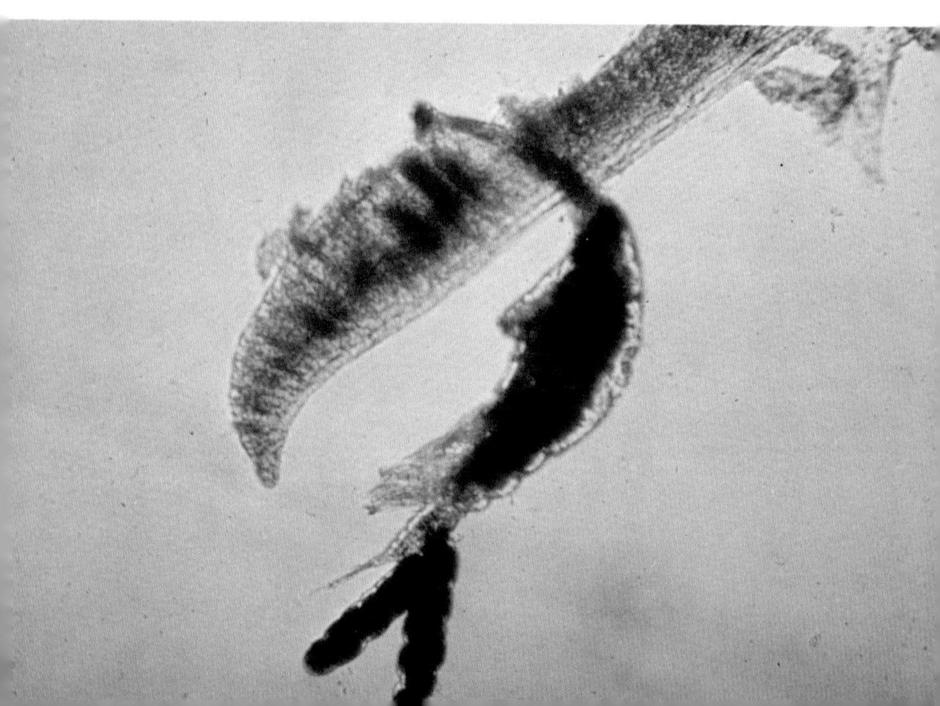

Tracheliastes maculatus in skin of *Leuciscus vulgaris.* Courtesy of P. de Kinkelin, Laboratoire d'Ichthyopathologie, Route de Thiverval, 78 Thiverval-Grignon, France.

Lernaea sp. Head region, greatly magnified. Photo by Frickhinger.

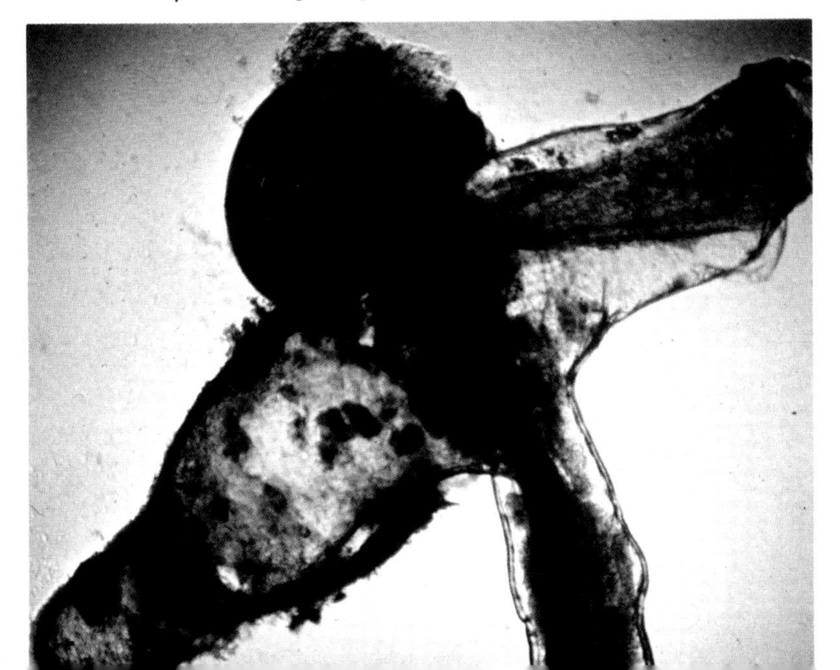

Table 8. Cestodes—*continued*

Parasite	Host	Treatment	Dosage	Method	Number of Appli-cations	Frequency	Author's Report of Success	Remarks	References
,, ,,	Unspecified	Dylox	0.8 ppm	I-M	1	....	Effective	Use to kill micro-crustaceons which serve as intermediate hosts	Funnikova, *et al.*, 1966
,, ,,	Unspecified	Kamala	2% of diet	0	7	Daily	Effective	....	Bauer, 1966
,, ,,	Unspecified	Phenasal	1 g/kg of fish	0	6	Daily	Effective	....	Muzykovski, 1968
,, ,,	Unspecified	Phenothiazine	Not given	0	?	?	Effective	....	Muzykovski, 1968
,, ,,	Unspecified	Ziram	1.0 ppm	I-M	1	....	Effective	Use to destroy *Cyclops* which serves as intermediate host	Gretillat, 1965

Table 9. NEMATODES

Parasite	Host	Treatment	Dosage	Method	Number of Appli- cations	Frequency	Author's Report of Success	Remarks	References
Contracaecum bidentatum	*Acipenser* sp.	Santonin	0.04 g/fish	0	?	?	Effective	Combine with sugar and animal fat	Agapova, 1957
Philometra lusii	*Cyprinus carpio*	Management	....	M	....	....	Effective	Do not rear carp of different ages in same pond	Bauer, 1966
Philometra sp.	*Cyprinus carpio*	Masoten	1–2 ppm	I	1	....	Effective	Required 10 days	Plate, 1970
Raphidascaris sp.	Unspecified	Drying	....	P	....	10 min.	Effective	Will kill eggs of nematodes	Engashev, 1966
Unidentified Nematode	Unspecified	Para-chlorometaxylenol	Not given, moisten food in drug	?	?	?	?	No details given	Amlacher, 1961b

Table 10. ACANTHOCEPHALANS

Parasite	Host	Treatment	Dosage	Method	Number of Appli- cations	Frequency	Author's Report of Success	Remarks	References
Echinorhynchus sp.	*Salmo gairdneri*	Bithionol	20,000 ppm in food	0	1	....	Effective	Removed 84% of infection	Gerard and de Kinkelin, 1971
Pomphorhynchus sp.	Unspecified	Di-N-butyl tin oxide	3.5 mg/kg of fish	0	3	Daily	Effective	....	Wellborn, Unpubl.

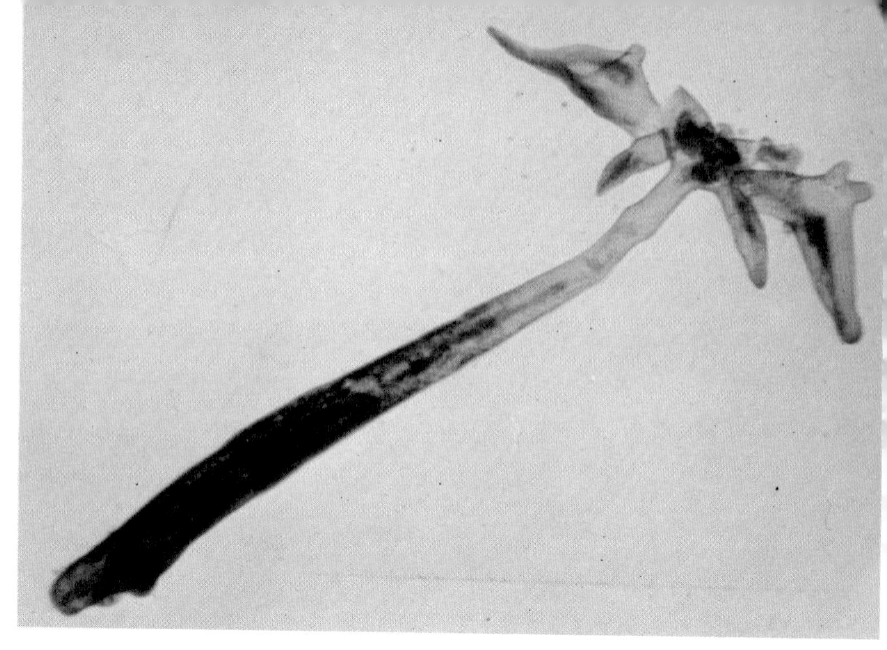

Lernaea cyprinacea, adult female removed from fish. Note 2 larger branched dorsal horns and 2 smaller ventral horns. Egg sacks missing. Photo by F. Meyer.

Lernaea cyprinacea, nauplius (free-swimming first larval stage). Photo by F. Meyer.

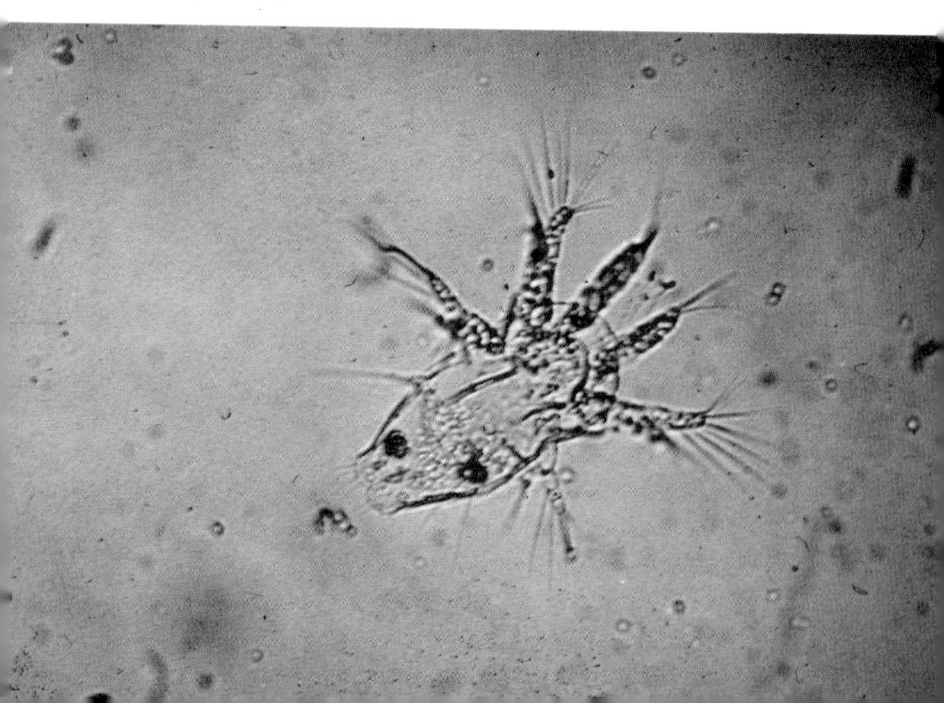

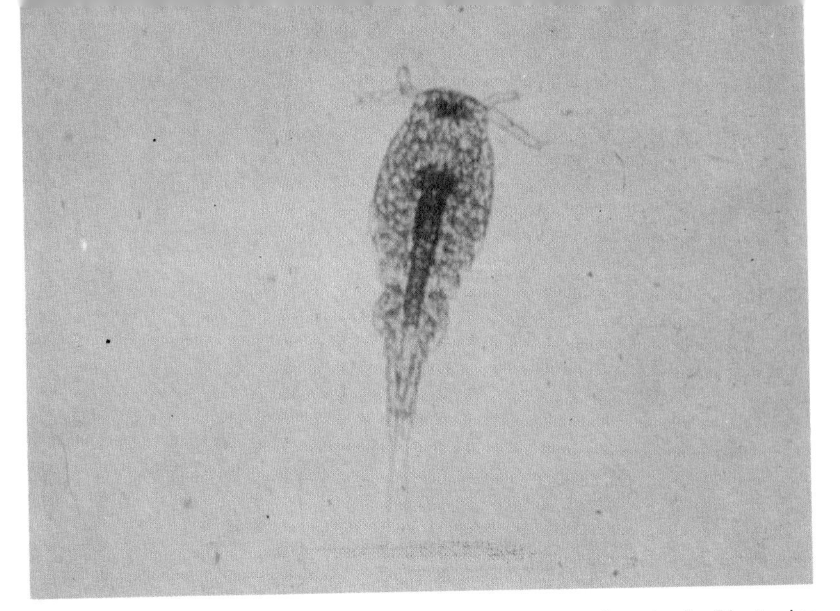

Lernaea cyprinacea, copepodid stage (free-swimming). Photo by F. Meyer.

Lernaea cyprinacea, copepodid stage attached to gill. Photo by F. Meyer.

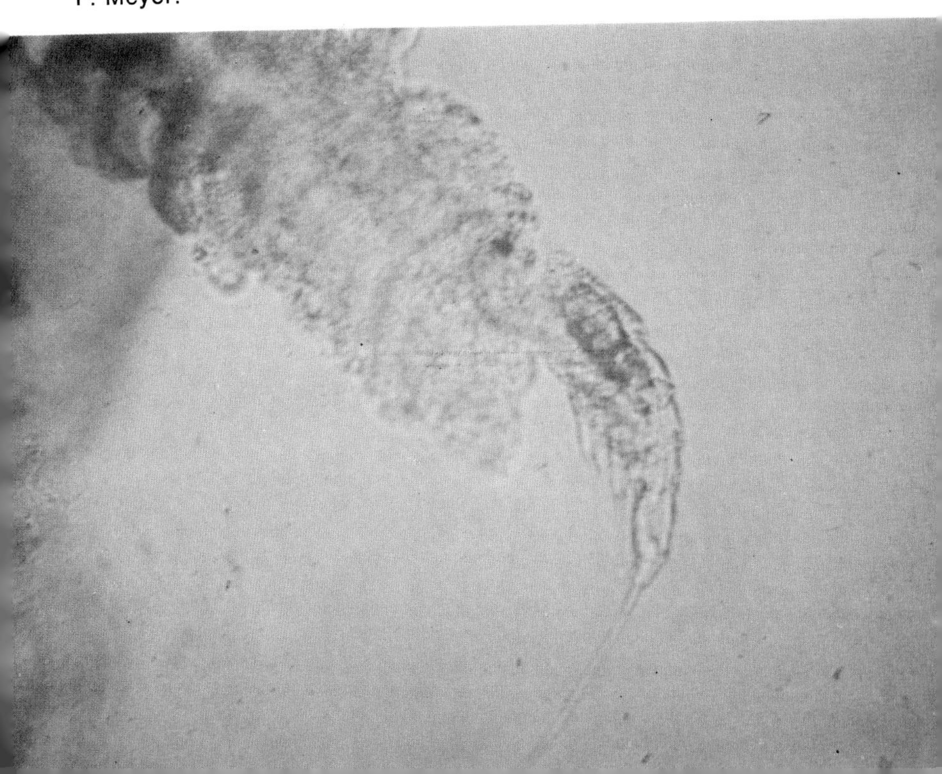

Table 11. LEECHES

Parasite	Host	Treatment	Dosage	Method	Number of Applications	Frequency	Author's Report of Success	Remarks	References
Erpobdella punctata	*In vitro*	Baygon	1.0 ppm	I	1	....	Not effective	Fish not included	F. Meyer, 1969a
" "	*In vitro*	Baytex	0.5 ppm	I	1	....	Effective	Fish not included	F. Meyer, 1969a
" "	*In vitro*	Dylox	0.5 ppm	I	1	....	Effective	Fish not included	F. Meyer, 1969a
Hemiclepsis marginata	*Labeo, Catla, Cirrhina*	Gammexane	0.5 ppm	I	1	....	Effective	....	Saha and Sen, 1955
Illinobdella moorei	*In vitro*	Baygon	0.5 ppm	I	1	....	Effective	Fish not included	F. Meyer, 1969a
" "	*In vitro*	Baytex	1.0 ppm	I	1	....	Not effective	Fish not included	F. Meyer, 1969a
" "	*In vitro*	Dylox	0.125 ppm	I	1	....	Effective	Fish not included	F. Meyer, 1969a
Piscicola geometra	*Cyprinus carpio*	Cupric chloride	5 ppm	F 15 min.	1	....	Effective	....	Bauer, 1958
" "	*Cyprinus carpio*	DDT	10,000 ppm	D ?	1	....	Effective	Toxic to many fish	Woynarovich, 1954
" "	*Cyprinus carpio*	Foschlor	0.75–1.0 ppm	I	1	....	Effective	....	Prost and Studnicka, 1968
" "	*Cyprinus carpio*	Lysol	2,000 ppm	D 5–15 sec.	1	....	Effective	....	Amlacher, 1961b; Schäperclaus, 1954; Bauer, 1958
" "	*Cyprinus carpio*	Priasol	400 ppm	D 5–15 sec.	1	....	Effective	....	Amlacher, 1961b
" "	Removed from fish	Quicklime	880–1,320 lbs/acre	I	1	....	Effective	To kill leeches in pond bottom	Amlacher, 1961b
" "	*Cyprinus carpio*	Quicklime	2,000 ppm	D 2–5 sec.	1	....	Effective	....	Schäperclaus, 1954; Bauer, 1958
" "	*Cyprinus carpio*	Sodium chloride	25,000 ppm	F 1 hr.	1	....	Effective	....	Bauer, 1958

[continued

Table 11. Leeches—*continued*

Parasite	Host	Treatment	Dosage	Method	Number of Applications	Frequency	Author's Report of Success	Remarks	References
Piscicola salmositica	*In vitro*	Baygon	1.0 ppm	I	1	....	Effective	Fish not included	F. Meyer, 1969a
,, ,,	*In vitro*	Baytex	1.0 ppm	I	1	....	Not effective	Fish not included	F. Meyer, 1969a
,, ,,	*In vitro*	Dylox	0.5 ppm	I	1	....	Effective	Fish not included	F. Meyer, 1969a
,, ,,	*Oncorhynchus gorbuscha*	Formalin	250 ppm	?	1	....	Not effective	....	Earp and Schwab, 1954
,, ,,	*Oncorhynchus gorbuscha*	PMA	2 ppm	F 1 hr.	1	....	Not effective	....	Earp and Schwab, 1954
,, ,,	*Oncorhynchus gorbuscha*	Roccal	2 ppm	F 1 hr.	1	....	Not effective	....	Earp and Schwab, 1954
,, ,,	*Oncorhynchus gorbuscha*	Sea water	100%	F 1 hr.	1	....	Effective	....	Earp and Schwab, 1954
Piscicola sp.	Unknown	Masoten	0.5 ppm	I	1	....	Effective	....	Plate, 1970
Placobdella parasitica	*In vitro*	Baygon	0.25 ppm	I	1	....	Effective	Fish not included	F. Meyer, 1969a
,, ,,	*In vitro*	Baytex	0.35 ppm	I	1	....	Effective	Fish not included	F. Meyer, 1969a
,, ,,	*In vitro*	Dylox	0.125 ppm	I	1	....	Effective	Fish not included	F. Meyer, 1969a
Theromyzon sp.	*In vitro*	Baygon	1.0 ppm	I	1	....	Not effective	Tested at 4.4°C	F. Meyer, 1969a
,, ,,	*In vitro*	Baytex	1.0 ppm	I	1	....	Not effective	Tested at 4.4°C	F. Meyer, 1969a
,, ,,	*In vitro*	Dylox	0.25 ppm	I	1	....	Effective	Tested at 4.4°C	F. Meyer, 1969a
Unidentified leeches	Pond fishes	Acetic acid, glacial	1,000 ppm	D	?	....	....	....	Khan, 1944
,, ,,	*Esox lucius*	Copper sulfate	0.5 ppm	F 5–6 hrs.	1	....	Effective	....	Avdosev, *et al.*, 1962

Sphaerid clams attached to mouths of rainbow trout. Courtesy of Dr. P. Ghittino, Istituto Zooprofilattico Sperimentale del Piemonte e della Liguria, Torino, Italy, and Edizioni Rivista di Zootecnia.

Glochidia (clam larvae) in gills of North American fish. Photo by G. Hoffman.

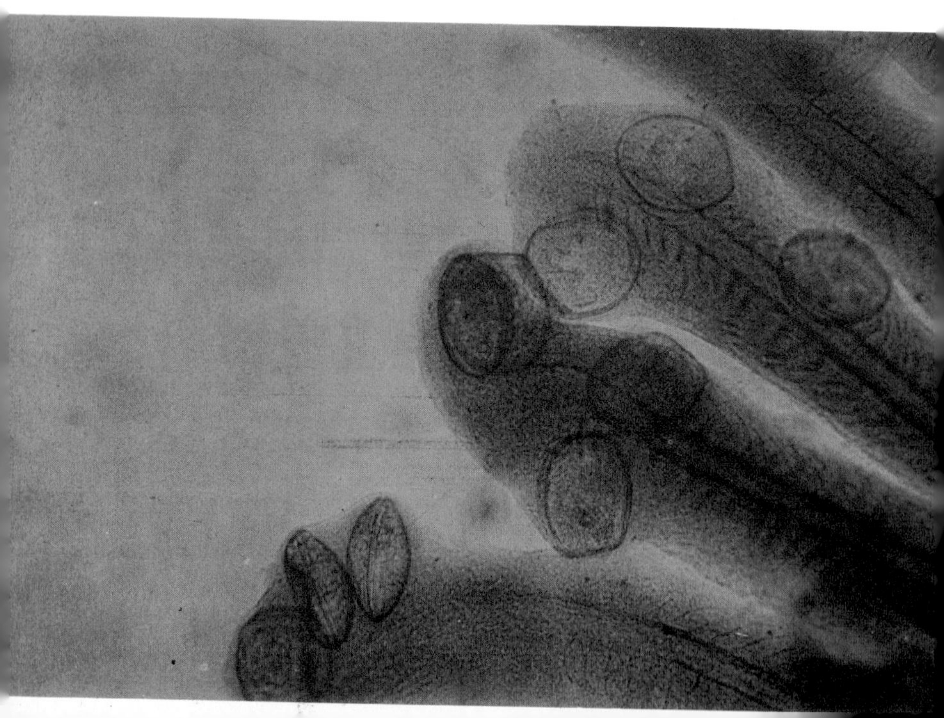

,,	,,	Unidentified	Copper sulfate	Not given	?	....	....	Not effective	Effective in aquariums but not in ponds	Moore, 1923
,,	,,	*Labeo, Catla, Cirrhina*	Gammexane	0.5 ppm	I	1	....	Effective	....	Saha and Sen, 1955
,,	,,	Unidentified	Lime	?	I	1	....	Effective	....	Quebec Game and Fisheries Dept., 1948
,,	,,	Unidentified	Nicotine sulfate	?	I	1	....	Not effective	Effective in aquariums but not in ponds	Moore, 1923
,,	,,	Pond fishes	Potassium permanganate	100 ppm	D	?	....	....	....	Khan, 1944
,,	,,	Pond fishes	Sodium chloride	25,000 ppm	F 30 min.	1	....	Effective	....	Gopalakrishnan, 1963
,,	,,	Unidentified	Sodium chloride	30,000 ppm	D 5–15 sec.	1	....	Effective	....	F. Meyer, 1969a
,,	,,	Unidentified	Zinc chloride	Not given	?	....	....	Not effective	Effective in aquariums but not in ponds	Moore, 1923

Table 12. PARASITIC COPEPODS

Parasite	Host	Treatment	Dosage	Method	Number of Appli-cations	Frequency	Author's Report of Success	Remarks	References
Achtheres micropteri	*Ictalurus punctatus*	Dylox	0.25 ppm	I	4	Weekly	Effective	....	F. Meyer, 1970
Achtheres sp.	Unidentified	DDT	0.25 ppm	I	?	?	Effective	....	Schäperclaus, 1954
,, ,,	Unidentified	Dylox	0.25 ppm	I	4	Weekly	Effective	....	Kimura, 1967
Argulus foliaceus	Unknown	Masoten	25,000 ppm	D 5–10 min.	1	....	Effective	....	Plate, 1970
,, ,,	*Salmo gairdneri, Anguilla anguilla*	Masoten	0.25–0.30 ppm	I	2	1–2 weeks	Effective	21–29°C	Ghittino and Arcarese, 1970
,, ,,	Unidentified	Sodium chloride	25,000 ppm	F 5 hrs.	1	....	Effective	Toxic to fish	Chen, 1933
Argulus japonicus	*Cyprinus carpio*	Lindane	100 ppm	F 30 min.	1	....	Effective	....	Kiselev and Ivleva, 1950, 1953
,, ,,	*Ctenopharyngodon idellus*	Dipterex	2 ppm	I	1	....	Effective	....	Kimura, 1967
Argulus sp.	Pond fishes	Acetic acid, glacial	1,000 ppm	D 5 min.	1	....	Uncertain	....	Khan, 1944
,, ,,	Unidentified	Ammonium chloride	500 ppm	I	1	....	Effective	Flush after 24 hrs.	Chen, 1933
,, ,,	Unidentified	Ammonium chloride	1,000 ppm	F 4 hrs.	1	....	Effective	....	Chen, 1933
,, ,,	*Cyprinus carpio*	Balsam of Peru Oil	40 ppm	F 3 hrs.	1	....	Effective	....	Kelly, 1962
,, ,,	Pond fishes	Bamboo slats	....	P	?	Weekly	Not effective	Fish rub off parasites	De, 1910; Southwell, 1915; and Alikunhi, 1957
,, ,,	*Carassius auratus*	Baytex	0.12 ppm	I	2	Triweekly	Effective	....	Osborn, 1966

,,	,,	*Cyprinus carpio*	Benzene hexachloride	0.12 ppm	I	1	....	Effective	Also killed larval stages	Lahav, Sarig and Shilo, 1966; Lahav and Sarig, 1967
,,	,,	Unidentified	Benzene hexachloride	0.078 ppm	I	1	....	Effective	....	Sproston, 1956
,,	,,	Pond fishes	Brackish water	....	I	?	....	Effective	....	Hora, 1943
,,	,,	*Cyprinus carpio*	Bromex-50	0.12 ppm	I	1	....	Effective	....	Lahav, Sarig and Shilo, 1966; Sarig, 1968
,,	,,	*Cyprinus carpio*	Copper sulfate	0.5 ppm + 0.2 ppm ferric sulfate	I	....	....	Not effective	....	Musselius and Strelkov, 1968
,,	,,	Aquarium fishes	DDFT (Gix)	2 drops/m^3	I	1	....	Effective	....	Meinken, 1954
,,	,,	*Cyprinus carpio*	DDFT (Gix)	100 ppm	D 5 min.	1	....	Effective	Leaves residue in fish flesh	Stammer, 1959
,,	,,	*Cyprinus carpio*	DDFT (Gix)	1 ppm	F 30 min.	1	....	Effective	....	Stammer, 1959
,,	,,	*Cyprinus carpio*	DDT	1,000 ppm	D 1 min.	1	....	Effective	Leaves residue in fish flesh	Bauer, 1959
,,	,,	*Labeo rohita*	DDT	0.005 ppm	F 24 hrs.	1	....	Effective	Leaves residue in fish flesh	Yousuf-Ali, 1968
,,	,,	Unidentified	DDT	0.01 ppm	I	?	....	Effective	Leaves residue in fish flesh	Schäperclaus, 1954
,,	,,	*Cyprinus carpio*	DDVP	0.25 ppm	I	1	....	Effective	....	Sarig, 1966 and 1968
,,	,,	Grass carp	Dipterex	2.00 ppm	I	?	?	Effective	....	Prowse, 1965

[*continued*

Table 12. Parasitic Copepods—*continued*

Parasite	Host	Treatment	Dosage	Method	Number of Appli-cations	Frequency	Author's Report of Success	Remarks	References
Argulus sp.	*Cyprinus carpio*	Dipterex	100 ppm	F 1 hr.	1	....	Effective	....	Sukhenko, 1963; Naumova, 1968
,, ,,	*Cyprinus carpio*	Dipterex	0.25 ppm	I	1	....	Effective	....	Sarig, 1966 and 1968
,, ,,	*Cyprinus carpio*	Drying	....	P	1	....	Effective	Dry pond thoroughly before restocking	Plehn, 1924; Loyen, 1931
,, ,,	Unidentified	Egg removal	....	M	....	Triweekly	....	Use stakes and wood floats to collect eggs of parasites	Kabata, 1970
,, ,,	Unspecified	Filters	....	P	....	Continuous	Inhibitory	Removes larvae and adults	Anon., 1960
,, ,,	Unspecified	Temperature	....	P	....	....	....	Freezing kills adults, not eggs	Loyen, 1931
,, ,,	Unidentified	*Gambusia* sp.	....	B	....	....	....	Prey on parasite larvae	Campbell, 1950
,, ,,	Pond fishes	Gammexane	0.2 ppm	I	2	Weekly	Effective	....	Saha and Sen, 1955
,, ,,	*Pylodictus olivaris*	*Lepomis* sp.; *Pomoxis* sp.	....	B	1	Continuous	Inhibitory	Not reliable, small fish eat *Argulus*	Spall, 1970
,, ,,	*Cyprinus carpio*	Lindane	0.0013 ppm	I	1	....	Effective	....	Hindle, 1949
,, ,,	Pond fishes	Lindane	0.112 ppm	I	1	....	Effective	....	Malacca Res. Inst., 1963
,, ,,	*Cyprinus carpio*	Lindane	0.5 ppm	I	1	....	Effective	....	Ivasik and Svirepo, 1964
,, ,,	*Cyprinus carpio*	Lindane	100 ppm	F 30 min.	1	....	Effective	....	Kiselev and Ivleva, 1950 and 1953

,,	,,	*Cyprinus carpio*	Lindane	0.02 ppm	I	1	....	Effective	....	Sarig, 1966 and 1968
,,	,,	*Cyprinus carpio*	Lindane	0.007 ppm	I	1	....	Effective	....	Stammer, 1959
,,	,,	*Cyprinus carpio*	Lindane	....	....	....	....	....	*Argulus* became partially resistant	Lahav and Shilo, 1962
,,	,,	*Cyprinus carpio*	Lysol	2,000 ppm	D 5–15 sec.	1	....	Effective	....	Bauer, 1958; Amlacher, 1961b
,,	,,	Unidentified	Lysol	2,000 ppm	D 5–15 sec.	1	....	Not effective	Removes parasites but does not kill them	Schäperclaus, 1954
,,	,,	*Cyprinus carpio*	Malathion	0.25 ppm	I	1	....	Effective	....	Sarig, 1966 and 1968
,,	,,	Unknown	Masoten	0.5 ppm	I	1	....	Effective	....	Plate, 1970
,,	,,	Unknown	Masoten	1.0 ppm	I	1	....	Effective	....	Plate, 1970
,,	,,	*Cyprinus carpio*	Methyl parathion	0.125 ppm	I	1	....	Effective	....	Bowen and Putz, 1966
,,	,,	Unidentified	Minnows	....	B	....	....	....	To prey on larvae of parasites	Hofer, 1904; Campbell, 1950
,,	,,	*Carassius auratus*, *Cyprinus carpio*	Neguvon	50,000 ppm	F 30 min.	1	....	Effective	....	Bailosoff, 1963
,,	,,	*Carassius auratus*, *Cyprinus carpio*	Neguvon	35,000 ppm	D 1 min.	1	....	Effective	....	Bailosoff, 1963
,,	,,	Tropical fishes	Neguvon	35,000 ppm	D 3 min.	1	....	Effective	....	Van Duijn, 1967
,,	,,	*Cyprinus carpio*	Oxygenation	....	M	....	....	Helpful	Increase aeration in pond	Bauer, 1959
,,	,,	*Cyprinus carpio*	Parathion	1.0 ppm	I	1	....	Effective	....	Stammer, 1959
,,	,,	*Cyprinus carpio*	Peru oil	40 ppm	F 3 hrs.	1	....	Effective	....	Kelly, 1962

[*continued*

Table 12. Parasitic Copepods—*continued*

Parasites	Host	Treatment	Dosage	Method	Number of Applications	Frequency	Author's Report of Success	Remarks	References
Argulus sp.	*Ctenopharyngodon idellus*	Potassium permanganate	10 ppm	....	....	....	Not effective	Killed fish before parasite	Chen, 1933
,, ,,	Unidentified	Potassium permanganate	100 ppm	F 5–10 min.	1	....	Effective	....	Brunner, 1943
,, ,,	Unidentified	Potassium permanganate	500 ppm	D 5 min.	1	....	Effective	....	Brunner, 1943
,, ,,	Unidentified	Potassium permanganate	1,000 ppm	D 2 min.	1	....	Effective	....	Brunner, 1943
,, ,,	Unidentified	Potassium permanganate	10 ppm	F 30 min.	1	....	Effective	....	Kiselev and Ivleva, 1950; Schäperclaus, 1954
,, ,,	*Cyprinus carpio*	Priasol	4,000 ppm	D 5–15 sec.	1	....	Effective	....	Amlacher, 1961b
,, ,,	*Cyprinus carpio*	Pyrethrum	0.01 ppm	F 50 min.	1	....	Effective	1% powder not a suitable formulation	Kemper, 1933; Stammer, 1959
,, ,,	*Cyprinus carpio*	Pyrethrum	0.1 ppm	F 20 min.	1	....	Effective	....	Stammer, 1959
,, ,,	*Cyprinus carpio*	Pyrethrum	10.0 ppm	F 3 min.	1	....	Effective	Toxic to fish	Stammer, 1959
,, ,,	*Cyprinus carpio*	Pyrethrum	20.0 ppm	F 1 min.	1	....	Effective	....	Bauer, 1961
,, ,,	Unidentified	Removal by hand picking	....	S	....	....	Effective	Only on limited numbers of fish	Hofer, 1904; Roth, 1922
,, ,,	Pond fishes	Sodium chloride	10,000 ppm	F 1 hr.	1	....	Effective	....	Khan, 1944
,, ,,	*Catla catla, Cirrhina mrigola, Labeo rohita, Tilapia mossambica*	Sodium chloride	30,000 ppm	D 5 min.	1	....	Effective	Only removes, does not kill parasites	Yousuf-Ali, 1968

,, ,,	*Tilapia mossambica*	Sodium chloride	Build up to 13,000 ppm	I	1	....	Effective	Build up in 3 days	Yousuf-Ali, 1968
,, ,,	Unidentified	Teaseed cake	6 ppm	I	1	....	Effective	Highly toxic to fish	Chen, 1933
Caligus spinosus	Unknown	Masoten	100 ppm	D 5 sec.	1	....	Effective	In sea water	Plate, 1970
Caligus sp.	Unidentified	Lindane	0.1 ppm	I	1	....	Effective	Toxic to some fish	DeGraaf, 1959
Ergasilus sieboldi	Unidentified	Hydrogen sulfide	?	I	....	....	Probably effective	Not thoroughly tested	Gnadeberg, 1949
,, ,,	Unidentified	Masoten	0.15 ppm	I	1	....	Effective	....	Plate, 1970
Ergasilus sp.	Unidentified	Benzene hexachloride	0.078 ppm	I	1	....	Not effective	....	Sproston, 1956
,, ,,	*Mugil capito, Mugil cephalus*	Bromex-50	0.3 ppm	I	4–6	Biweekly	Effective	....	Sarig, 1968
,, ,,	*Mugil capito, Mugil cephalus*	Bromex-50	0.1 ppm	I	1	....	Effective	....	Sarig, 1969
,, ,,	Unidentified	DDT	0.01	I	1	?	Effective	May leave residue in fish flesh	Schäperclaus, 1954
,, ,,	*Ctenopharyngodon idellus*	Dipterex	2.0 ppm	I	?	?	Effective	....	Prowse, 1965
,, ,,	*Cyprinus carpio*	Dylox	0.15 ppm	I	1	....	Effective	....	Lahav and Sarig, 1967
,, ,,	Unspecified	Filters	....	P	....	Continuous	Inhibitory	Remove larvae	Anon., 1968
,, ,,	*Mugil* sp.	Malathion	0.20 ppm	I	1	?	Effective	....	Lahav and Sarig, 1967
,, ,,	Unidentified	Gesarol	20,000	F 30 min.	1	?	Effective	....	Schäperclaus, 1954

[continued

Table 12. Parasitic Copepods—*continued*

Parasite	Host	Treatment	Dosage	Method	Number of Appli-cations	Frequency	Author's Report of Success	Remarks	References
Lernaea cyprinacea	Lake fishes	Acidic water	....	....	....	Continuous	....	No copepod parasites found in acidic lakes	Bere, 1935
,, ,,	*Notemigonus crysoleucas*	Antimycin-A	0.005 ppm	I	4	Weekly	Not effective	....	F. Meyer, Unpubl.
,, ,,	*Notemigonus crysoleucas*	Baygon	1.0 ppm	I	4	Weekly	Not effective	Inhibited	F. Meyer, Unpubl.
,, ,,	*Notemigonus crysoleucas*	Baytex	0.25 ppm	I	4	Weekly	Not effective	....	F. Meyer, Unpubl.
,, ,,	*Notemigonus crysoleucas*	Benzene hexachloride	0.0625 ppm	I	4	Weekly	Effective	Toxic reactions by fish	F. Meyer, 1966c
,, ,,	*Carassius auratus*	Benzene hexachloride	0.0625 ppm	I	6	5 days	Effective	....	Giudice, 1950
,, ,,	*Notemigonus crysoleucas*	Bromex-50	1.0 ppm	I	4	Weekly	Not effective	....	F. Meyer, Unpubl.
,, ,,	*Notemigonus crysoleucas*	Bromex-50	4.0 ppm	I	4	Weekly	Not effective	Toxic to fish	F. Meyer, Unpubl.
,, ,,	*Lepomis macrochirus*	Bromex-50	2.0 ppm	I	1	....	....	Toxic to fish	F. Meyer, Unpubl.
,, ,,	*Ictalurus punctatus*	Bromex-50	3.2 ppm	I	1	....	....	Toxic to fish	F. Meyer, Unpubl.
,, ,,	*Notemigonus crysoleucas*	7-Co-ral	0.4% of diet	O	....	Daily	Not effective	Toxic to fish	F. Meyer, Unpubl.

,,	,,	*Cyprinus carpio, Carassius auratus*	7-Co-ral	0.1 ppm	I	4	Weekly	Not effective	....	F. Meyer, Unpubl.
,,	,,	*Ictalurus punctatus, Lepomis macrochirus, Micropterus salmoides*	7-Co-ral	0.1 ppm	I	4	Weekly	Not effective	....	F. Meyer, Unpubl.
,,	,,	*Notemigonus crysoleucas*	Dibrom	0.5 ppm	I	4	Weekly	Inhibitory	....	F. Meyer, Unpubl.
,,	,,	*Notemigonus crysoleucas*	Dibrom	1.0 ppm	I	4	Weekly	Inhibitory	....	F. Meyer, Unpubl.
,,	,,	Pond fishes	Dipterex	0.2–0.5 ppm	I	3	3 weeks	Effective	Excellent control of larval forms	Kasahara, 1962, 1968
,,	,,	*Cyprinus carpio*	Dipterex	0.25 ppm	I	Few	Within 20 days	Effective	Excellent control of larval forms	Sarig, 1966 and 1968
,,	,,	*Notemigonus crysoleucas*	Dursban	0.05 ppm	I	5	Weekly	....	Toxic to fish	F. Meyer, 1969b
,,	,,	*Notemigonus crysoleucas*	Dursban	0.02 ppm	I	5	Weekly	Effective	Caused scoliosis	F. Meyer, 1969b
,,	,,	*Carassius auratus, Cyprinus carpio, Ictalurus punctatus, Lepomis macrochirus, Micropterus salmoides, Notemigonus crysgleucas*	Dylox	0.25 ppm	I	4	Weekly	Effective	Excellent results	F. Meyer, 1966c and 1970; F. Meyer, Unpubl.

[continued

Table 12. Parasitic Copepods—*continued*

Parasite	Host	Treatment	Dosage	Method	Number of Appli-cations	Frequency	Author's Report of Success	Remarks	References
Lernaea cyprinacea	*Notemigonus crysoleucas*	Dylox	0.25% in diet	0	....	Daily	Inhibitory	....	F. Meyer, Unpubl.
,, ,,	*Carassius auratus*	Dylox	20.0 ppm	D 30 min.	1	....	Not effective	Toxic to some fish	F. Meyer, Unpubl.
,, ,,	*Carassius auratus*	Dylox	0.2 ppm	I	2	2 weeks	Effective	....	R. Allison, 1969
,, ,,	Unspecified	Filters	....	P	....	Continuous	Inhibitory	Removes larvae	Anon., 1968
,, ,,	*Catla* sp., *Cyprinus carpio, Ictalurus punctatus, Lepomis macrochirus, Micropterus Salmoides*	Gammexane	1.0 ppm	I	1	....	Effective	....	Saha, *et al.*, 1959
,, ,,	*Notemigonus crysoleucas*	Korlan	0.25 ppm	I	4	Weekly	Inhibitory	Higher levels toxic to fish	F. Meyer, Unpubl.
,, ,,	*Notemigonus crysoleucas*	Korlan	0.25% in diet	0	....	Daily	Inhibitory	Toxic to fish	F. Meyer, Unpubl.
,, ,,	*Carassius auratus*	Korlan	0.25 ppm	I	4	Weekly	Inhibitory	Toxic to fish	F. Meyer, Unpubl.
,, ,,	*Carassius auratus*	Lexone (contains BHC)	2.66 ppm	I	?	Weekly	....	Reduced parasites	Giudice, 1950
,, ,,	*Salmo gairdneri*	Lindane	100.0 ppm	F 1 hr.	1	....	Inhibitory	Toxic to fish	McNeil, 1961
,, ,,	*Notemigonus crysoleucas*	Lindane	....	....	....	....	....	Toxic in reccomended conc.	Lewis, 1961
,, ,,	*Salmo gairdneri*	Masoten	0.25–0.30 ppm	I	2	Weekly	Effective	?1–24°C	Ghittino and Arcarese, 1970

,,	,,	*Notemigonus crysoleucas*	Menazon	0.4% in diet	O	....	Daily	Not effective	Toxic to fry	F. Meyer, Unpubl.
,,	,,	Pond fishes	*Mesocyclops leuckarti*	....	B	1	Continuous	Inhibitory	Eats larval *Lernaea*	Kasahara, 1962
,,	,,	*Notemigonus crysoleucas*	Methyl parathion	0.125 ppm	I	4	Weekly	Not effective	....	F. Meyer, Unpubl.
,,	,,	*Notemigonus crysoleucas*	Mitox	20.0 ppm	I	4	Weekly	Not effective	....	F. Meyer, Unpubl.
,,	,,	*Notemigonus crysoleucas*	Naled	1.0 ppm	I	4	Weekly	Not effective	....	F. Meyer, 1969b
,,	,,	Unspecified	Potassium permanganate	20.0 ppm	D 90 min.	1	....	Inhibitory	Use at 15-20°C	Yui-fan, *et al.*, 1961; Yin Wen-Ying, *et al.*, 1963
,,	,,	?	Potassium permanganate	21.6 ppm	?	?	?	Effective	Toxic to most fish	Fletcher, 1961
,,	,,	*Cyprinus carpio*	Potassium permanganate	25 ppm	F $1\frac{1}{2}$ hrs.	?	....	Effective	....	Sarig, 1966
,,	,,	Unspecified	Potassium permanganate	10.0 ppm	D 90 min.	1	....	Inhibitory	Use at 21–30°C	Yin Wen-Ying, *et al.*, 1963
,,	,,	Pond fishes	Potassium permanganate	50,000 ppm	D	1	....	Inhibitory	....	Gopalakrishnan, 1963
,,	,,	Pond fishes	Potassium permanganate	Concentrated (?)	T	?	As needed	Effective	....	Van Duijn, 1967
,,	,,	Unspecified	Pyrethrum	30–100 ppm	F 10–20 min.	?	?	Effective	....	Kemper, 1933; Dogiel, *et al.*, 1958
,,	,,	*Notemigonus crysoleucas*	Ronnel	0.25 ppm	I	4	Weekly	Not effective	Toxic to fry	F. Meyer, Unpubl.

[continued

Table 12. Parasitic Copepods—*continued*

Parasite	Host	Treatment	Dosage	Method	Number of Applications	Frequency	Author's Report of Success	Remarks	References
Lernaea cyprinacea	*Notemigonus crysoleucas*	Ronnel	0.8% in diet	O	....	Daily	Not effective	Toxic to fry	F. Meyer, Unpubl.
,, ,,	*Notemigonus crysoleucas*	Ruelene	0.1% in diet	O	....	Daily	Inhibitory	Higher levels toxic to fish	F. Meyer, Unpubl.
,, ,,	*Carassius auratus, Cyprinus carpio, Ictalurus punctatus, Lepomis macrochirus, Micropterus salmoides, Notemigonus crysoleucas*	Ruelene	0.25 ppm	I	4	Weekly	Not effective	....	F. Meyer, Unpubl.
,, ,,	Pond fishes	Sodium chloride	20,000 ppm	F 15 min.	?	?	....	....	Nakai and Kokai, 1931
,, ,,	*Notemigonus crysoleucas*	Tiguvon	0.6% in diet	O	....	Daily	Not effective	....	F. Meyer, Unpubl.
,, ,,	*Notemigonus crysoleucas*	Zectran	1.0 ppm	I	4	Weekly	Effective	Toxic to fry	F. Meyer, Unpubl.
Lernaea sp.	*Slamo gairdneri*	Benzene hexachloride	0.062 ppm	I	6	5 days	Effective	....	McNeil, 1961
,, ,,	*Cyprinus carpio*	Benzene hexachloride	0.12 ppm	I	?	Weekly	Effective	Killed larval parasites	Lahav, Sarig and Shilo, 1964; Lahav and Sarig, 1967

,,	,,	Unidentified	Benzene hexachloride	0.078 ppm	I	?	?	Not effective	....	Sproston, 1956
,,	,,	*Cyprinus carpio*	Bromex-50	0.1 ppm	I	4	Weekly	Effective	....	Lahav, Sarig and Shilo, 1966
,,	,,	*Cyprinus carpio*	Bromex-50	0.12 ppm	I	4	Weekly	Effective	....	Sarig, 1968
,,	,,	Unidentified	Calcium hypochlorite	1.0 ppm (as free chlorine)	I	?	Triweekly	Inhibitory	Toxic to many fish	Schäperclaus, 1954
,,	,,	Unidentified	Castor bean plant	Bundles	I	?	?	....	Reported as toxic to *Lernaea*	Hickling, 1962
,,	,,	Unidentified	Chlorine	1.0 ppm	I	?	3 days	Inhibitory	Toxic to many fish	Putz and Bowen, 1964
,,	,,	Unidentified	DDT	20,000 ppm	F 30 min.	1	....	Inhibitory	....	Schäperclaus, 1954
,,	,,	Unidentified	DDT	0.02 ppm	I	1	....	Inhibitory	Toxic to crayfish	Schäperclaus, 1954
,,	,,	Unidentified	DDT	"Strong solution"	T	1	....	Effective	Apply to parasite	Van Duijn, 1967
,,	,,	*Cyprinus carpio*	DDVP	0.225 ppm	I	4	Weekly	Effective	....	Kabata, 1970
,,	,,	*Cyprinus carpio*	Dipterex	0.25 ppm	I	4	Weekly	Effective	....	Lahav, Sarig and Shilo, 1964
,,	,,	*Ctenopharyngodon idellus*	Dipterex	0.25 ppm	I	1	....	Effective	Killed copepodid stage of parasite	Prowse, 1965
,,	,,	Unidentified	Drying	....	P	....	....	Effective	....	Kabata, 1970
,,	,,	*Cyprinus carpio*	Dylox	0.5 ppm	I	5	Weekly	Effective	....	Kasahara, 1962
,,	,,	*Carassius auratus*	Dylox	0.25 ppm	I	4–6	Weekly	Effective	....	Osborn, 1966; Rogers, 1966
,,	,,	*Carassius auratus*	Dylox	0.5 ppm	I	2	Triweekly	Inhibitory	Kills copepodid stage of parasite	Osborn, 1966

[continued

Table 12. Parasitic Copepods—*continued*

Parasite	Host	Treatment	Dosage	Method	Number of Appli-cations	Frequency	Author's Report of Success	Remarks	References
Lernaea sp.	Unidentified	Dylox	0.25 ppm	I	5	Weekly	Effective	Killed larval stage of parasites	Musselius, 1967
,, ,,	Unidentified	Formalin	250 ppm	F 30 min.	1	....	Inhibitory	....	Putz and Bowen, 1964
,, ,,	*Salmo gairdneri*	Formalin	250 ppm	F 1 hr.	1	....	Poor	Toxic to fish	McNeil, 1961
,, ,,	*Cyprinus carpio*	Gammexane	0.2 ppm	I	?	?	Effective	....	Saha and Sen, 1955
,, ,,	*Lepomis macrochirus*	Gammexane	0.2 ppm	I	?	?	Poor	Toxic to fish	Snow, 1958
,, ,,	Unidentified	Lexone	0.27 ppm	I	?	Weekly	Effective	....	Putz and Bowen, 1964
,, ,,	Unidentified	Lindane	0.2 ppm	I	3	Weekly	Effective	....	Gopalakrishnan, 1964
,, ,,	Unidentified	Lindane	0.062 ppm	I	2	3rd day	Effective	....	Putz and Bowen, 1964
,, ,,	Unidentified	Malathion	1.0 ppm	I	?	?	....	....	Kabata, 1970
,, ,,	Unknown	Masoten	0.2–0.4 ppm	I	4	Weekly	Effective	....	Plate, 1970
,, ,,	Unidentified	*Mesocyclops*	....	B	1	?	....	Preys on larval *Lernaea*	Kasahara, 1962
,, ,,	Unidentified	Neguvon	0.2 ppm	I	?	?	....	....	Kabata, 1970
,, ,,	*Cyprinus carpio*	Potassium permanganate	25 ppm	F 25 min.	2	90 min	Effective	Toxic to many fish	Shilo, Sarig, and Rosenberg, 1960
,, ,,	*Cyprinus carpio*	Potassium permanganate	10 ppm	F 50 min.	1	....	Poor	Toxic to fish	Lahav, Sarig and Shilo, 1964

,, ,,	Unidentified	Potassium permanganate	10,000 ppm	T	1	....	....	Apply only to parasite	Van Duijn, 1967
,, ,,	Unidentified	Sodium chloride	13,000 ppm	I	4	Daily	Poor	Kills only nauplii, not effective below 14°C. Build up to 13,000 ppm in 3 days	Van Duijn, 1967
,, ,,	*Cyprinus carpio*	Sodium chloride	25,000 ppm	F 30 min.	....	?	Temporary	Only stuns parasite	Shilo, Sarig and Rosenberg, 1960
Salmincola sp.	*In vitro*	Acetic acid	10,000 ppm	....	....	....	Effective	Killed larvae but not adult parasites	Fasten, 1912
,, ,,	*In vitro*	Calcium chloride	8,500 ppm	....	....	....	Effective	Killed larvae but not adult parasites	Fasten, 1912
,, ,,	*In vitro*	Copper sulfate	2,000 ppm	....	....	....	Effective	Killed larvae but not adult parasites	Fasten, 1912
,, ,,	Salmonid fishes	Formalin	166–250 ppm	F 1 hr.	1	....	Effective	Killed larvae only	Kabata, 1970
,, ,,	Salmonid fishes	*Gambusia*	....	B	....	....	....	Feed on larval parasites	Kabata, 1970
,, ,,	*In vitro*	Hydrochloric acid	800 ppm	....	....	....	Effective	Killed larvae only	Fasten, 1912
,, ,,	Salmonid fishes	Lights	....	....	....	....	....	Use lights to attract larval parasites, then dip with small mesh nets	Kabata, 1970
,, ,,	*In vitro*	Magnesium sulfate	15,000 ppm	....	....	....	Effective	Killed larvae only	Fasten, 1912
,, ,,	*In vitro*	Nitric acid	300 ppm	....	....	....	Effective	Killed larvae only	Fasten, 1912
,, ,,	Salmonid fishes	*Notropis* sp.	....	B	....	....	....	Feed on larval parasites	Kabata, 1970

[*continued*

Table 12. Parasitic Copepods—*continued*

Parasite	Host	Treatment	Dosage	Method	Number of Appli- cations	Frequency	Author's Report of Success	Remarks	References
Salmincola sp.	*In vitro*	Oxalic acid	3,000 ppm	....	....	....	....	Killed larvae only	Fasten, 1912
,, ,,	*In vitro*	Potassium chlorate	2,000 ppm	....	....	....	....	Killed larvae only	Fasten, 1912
,, ,,	Salmonid fishes	Sand filters	....	P	....	....	Effective	Removes larvae from water supply	Fasten, 1912; Kabata, 1970
,, ,,	*In vitro*	Sodium chloride	20,000 ppm	....	....	....	Effective	Killed larvae only	Fasten, 1912
,, ,,	*In vitro*	Sulfuric acid	150 ppm	....	....	....	Effective	Killed larvae only	Fasten, 1912
,, ,,	*In vitro*	Tartaric acid	4,500 ppm	....	....	....	Effective	Killed larvae only	Fasten, 1912
Sinergasilus major	*Ctenopharyngodon idellus*	Chlorophos	400 ppm	I	1	....	Not effective	....	Bauer and Babaev, 1964
,, ,,	*Ctenopharyngodon idellus*	Chlorophos	100 ppm	F 3 hrs.	1	....	Effective	Toxic to fish	Bauer and Babaev, 1964
,, ,,	*Ctenopharyngodon idellus*	Copper sulfate + ferric sulfate (5:2)	0.7 ppm	I	1	....	Effective	....	Hsu Me-Keng, and Jen Jung-feng, 1955
Sinergasilus sp.	Unidentified	Copper sulfate + ferric sulfate (5:2)	0.7 ppm	I	1	....	Not effective	....	Musselius and Strelkov, 1968
,, ,,	Unidentified	Dylox	100 ppm	F 3 hrs.	1	....	Effective	Killed some fish	Bauer and Babaev, 1964
Tracheliastes sp.	Unidentified	DDT	0.02 ppm	I	?	....	Effective	Toxic to crayfish	Schäperclaus, 1954
Unidentified Copepod	*Hippocampus* sp.	Anthium dioxide	0.2 ppm	I	1	....	Effective	....	Zeiller, 1966

Table 13. MISCELLANEOUS, Turbellarian

Parasite	Host	Treatment	Dosage	Method	Number of Applications	Frequency	Author's Report of Success	Remarks	References
Planaria sp.	Unidentified	Ammonium nitrate	1,200 ppm	I	?	?	Effective	4–8°C	Schäperclaus, 1954
,, ,,	Unidentified	Acetic acid	2 Tablespoons in 25 liters of water	D	?	....	Effective	....	Reichenbach-Klinke, and Elkan, 1965
Unidentified Turbellaria	Aquarium fishes	Chloramin	66 ppm	F 2–4 hrs.	1	....	Effective	....	Reichenback-Klinke, 1966
,, ,,	Pond fishes	Chloramin	66 ppm	F 2–4 hrs.	1	....	Effective	....	Goncharov, 1966

Table 14. MISCELLANEOUS, Molluscs

Parasite	Host	Treatment	Dosage	Method	Number of Applications	Frequency	Author's Report of Success	Remarks	References
Australorbis glabratus	....	Acrolein	3 ppm	I	1	....	Effective	Toxic to fish	Ferguson, *et al.*, 1961
,, ,,	....	*Marisa cornuarietis*	....	B	....	....	Partially effective	Preys on *Australorbis* and other snails	Radke, Ritchie, and Ferguson, 1961
,, ,,	....	n-Tritylmorpholine	0.01–0.5 ppm	F 1 hr.	1	....	Effective	Higher concentration toxic to fish	Boyce, *et al.*, 1967; Crossland, 1967
,, ,,	....	n-Tritylmorpholine	0.01–0.05 ppm	F 24 hrs.	1	....	Effective	Higher concentration toxic to fish	Boyce, *et al.*, 1967; Crossland, 1967
,, ,,	....	n-Tritylmorpholine	0.025 ppm	I	1	....	Effective	Not toxic to fish	Boyce, *et al.*, 1967; Crossland, 1967
,, ,,	....	Trifluoromethyl nitrophenol	9 ppm	I	1	....	Effective	Toxic to many fish	Jobin and Unrau, 1967
Bulinus sp.	....	p-Nitrophenacyl chloride	2.5 ppm	I	1	....	Effective	Did not kill *Tilapia* sp.	Villiers and Mackenzie, 1963
Lymnaea peregra	*Salmo gairdneri*	Frescon (n-Tritylmorpholine)	0.1 ppm	I	?	5 weeks ?	Good temporary control	To control snail vector of *Diplostomulum spathaceum*	Crossland, *et al.*, 1971
Radix sp.	....	Nitrophenyl amidineura	0.05 ppm	I	1	....	Effective	Fish tolerated up to 70 ppm	Venulet and Schultz, 1964

Unidentified snails	Barium carbonate	?	I	1	....	Effective	Toxic to fish, persistent for several months	Deschiens, 1961
,, ,, ...	Barium chloride	?	I	1	....	Effective	Toxic to fish	Deschiens, 1961
,, ,, ...	Bayluscide	0.5–1.0 ppm	I	1	....	Effective	Toxic to fish	Deufel, 1964
,, ,, ...	Butyl tin oxide	1.0	I	1	....	Effective	Toxic to fry but not adult largemouth bass	B. F. Goodrich Co., 1966
,, ,, ...	Cadminum sulfate	5–15 ppm	I	1	....	Effective	Not toxic to *Carassius auratus, Cambarus,* or tadpoles	Deschiens and Tahiri, 1961
,, ,, ...	Calcium carbonate	195–1,135 kg/hectare (700–1,000 lb/A.)	I	1	....	Effective	Apply while pond is drained, flush before refilling	Amlacher, 1961b; Snow, 1962
,, ,, ...	Calcium oxide plus calcium hypochlorite (3:1)	2,385 kg/hectare (2,100 lb/A.)	I	1	....	Effective	Apply to drained pond, flush before refilling	Chechina, 1959
, ,, ...	Chevreul's salt	2.25 ppm	I	1	....	Effective	Did not kill fish	Deschiens, *et al.*, 1963; Gamet, *et al.*, 1964
,, ,, ...	Chlorophos	Unknown	?	?	?	?	....	Musselius and Laptev, 1967
,, ,, ...	Copper carbonate	9.76 g/m^2 (2 lbs/1,000 ft^2)	I	1	....	Effective	Use only if methyl orange alkalinity is less than 50 ppm	Mackenthun, 1958; Barbosa, 1961

[continued

Table 14. Miscellaneous Molluscs—*continued*

Parasite	Host	Treatment	Dosage	Method	Number of Appli- cations	Frequency	Author's Report of Success	Remarks	References
Unidentified snails	....	Copper sulfate	5 ppm	I	1	....	Effective	If water is less than 50 ppm in $CaCO_3$, use with caution	Ivasik, Stryzhak, and Turkevich, 1968
,, ,,	....	Copper sulfate + copper carbonate (2:1)	3 lbs/1,000 ft^2	I	1	....	Effective	Use only in waters with methyl orange alkalinity above 50 ppm	Mackenthun, 1958
,, ,,	....	Cupric sulfate	?	?	?	....	....	....	Batte, *et al.,* 1961
,, ,,	....	Cuprous chloride	2 ppm	I	1	....	Effective	....	Deschiens and Floch, 1964; Chabaud, Deschiens, and Le Corroller, 1965; Deschiens, *et al.,* 1965
,, ,,	....	Cuprous chloride plus potassium chloride	1.4 ppm	I	1	....	Effective	Effective	Floch, *et al.,* 1964
,, ,,	....	Cuprous chloride plus potassium chloride (5:6)	20 ppm	I	1	....	Effective	Did not kill *Lebistes reticulatus*	Floch, *et al.,* 1964
,, ,,	....	Cuprous oxide	5 ppm	I	1	....	Effective	Did not kill fish fry	Deschiens and Floch, 1964
,, ,,	....	Cuprous oxide	50 ppm	I	1	....	Effective	Toxic to fry but not nanoplankton	Deschiens and Floch, 1964

,,	,,	....	Cuprous oxide	2.7 ppm	I	1	....	Effective	....	Frick *et al.*, 1964
,,	,,	....	Cuprous oxide	16 ppm	I	1	....	Effective	Did not kill *Micropterus salmoides, Lepomis microlophus Lepomis auritus*	Ritchie, 1969
,,	,,	....	Cupric sulfate	1 ppm	I	1	....	Effective	Fish not killed	Batte, *et al.*, 1961
,,	,,	....	Dowco 212	0.25 ppm	I	1	....	Effective	Did not kill *Lebistes reticulatus*	Ehrenford, 1968
,,	,,	....	Gramoxone	2–10 ppm	I	1	....	Effective	Killed algae but not fish	Paulini, 1965; Paulini and Camey, 1965
,,	,,	....	Gramoxone	0.5 ppm	I	1	....	Effective	10 ppm did not kill fish	Camey, *et al.*, 1966; Paulini and Camey, 1965
,,	,,	....	*Helobdella punctatolineata*	....	B	1	....	Effective in lab.	Prey on snails so should be tried	McAnnaly and Moore, 1966
,,	,,	....	*Lepomis microlophus*	....	B	1	....	Partially effective	Prey on snails	Avault and Allison, 1965; Carothers and Allison, 1966
,,	,,	....	Nicotine (as sulfate)	12–15 kg/hectare	I	1	....	Effective	Apply to pond bottom before filling	Chemagro Corp., 1971
,,	,,	....	Quicklime	1,000–1,500 kg/ hectare 880–1,320 lbs/A.)	I	1	....	Effective	Kills snails in bottom mud	Amlacher, 1961b
,,	,,	....	Saponin	15–18 kg/hectare	I	1	....	Effective	Apply to pond bottom before filling, toxic to fish	Chemagro Corp., 1971

[continued

A group of guppies heavily infested with *Anodonta* glochidia. Photo by Dr. H. Schneider.

Table 14. Miscellaneous Molluscs—*continued*

Parasites	Host	Treatment	Dosage	Method	Number of Applications	Frequency	Author's Report of Success	Remarks	References
Unidentified snails	....	Sodium pentachlorophenate	5 ppm	I	1	....	Effective	Toxic to fish, degrades in 10 days	Osborn, 1966
,, ,,	....	Trifluoromethyl-nitrophenol	12 ppm	I	1	....	Partially effective	Toxic to *Pimephales promelas* and *Ictalurus punctatus*	F. Meyer, Unpubl.
,, ,,	....	n-Tritylmorpholine	0.01–0.05 ppm	F 24 hrs.	1	....	Effective	Some fish loss at higher concentrations	Boyce, *et al.*, 1967
,, ,,	....	n-Tritylmorpholine	0.025 ppm	F 16 days	1	....	Effective	No fish loss	Crossland, 1967
,, ,,	....	Zectran	7 ppm	I	1	....	Effective	Snails killed but fish tolerance is only 6–12 ppm	F. Meyer, Unpubl.
Unidentified glochidia	Trout	Sand filters	....	P	1	Continuous	Effective	....	Davis, 1953
,, ,,	Trout	Screen filters	....	P	1	Continuous	Effective	200 meshes per inch	Locke, 1963

TOXICITY OF PARASITICIDES

While the toxicity of a chemical usually varies according to the pH, hardness, temperature, and dissolved constituents in the water under investigation, it was deemed useful to provide the reader with some data concerning the relative toxicity of each compound. The authors acknowledge that the list is incomplete but our purpose is to provide some background information rather than a definitive report on the toxic properties of the various chemicals. The reader is reminded that strains of fish, sexes, ages, and development stages are likely to vary greatly in their susceptibility to a given toxicant.

In the following table, toxicity levels are reported as LC_0 = not lethal at this concentration; LC_{25} = concentration killed 25% of the population; LC_{100} = concentration killed all test fish. Chemicals were listed as "toxic" or "not toxic" if authors failed to give toxicity levels.

Table 15. KNOWN TOXICITY LIMITS OF SELECTED CHEMICALS USED IN FISH PARASITE CONTROL

Compound	Fish Species	Dosage—ppm	Toxicity Limits	Exposure Time	Temperature	References
Acetic Acid	*Ictalurus punctatus*	629	LC_0	1 hr.	25°C	Clemens and Sneed, 1959
,, ,,	*Ictalurus punctatus*	15.8	LC_0	72 hrs.	25°C	Clemens and Sneed, 1959
Acraflavine, neutral	*Ictalurus punctatus*	56.2	LC_0	1 hr.	20°C	Clemens and Sneed, 1959
,, ,,	*Ictalurus punctatus*	27.0	LC_1	48 hrs.	17°C	Willford, 1967a
,, ,,	*Ictalurus punctatus*	15.8	LC_0	72 hrs.	25°C	Clemens and Sneed, 1959
,, ,,	*Ictalurus punctatus*	4.2	LC_0	96 hrs.	20°C	Clemens and Sneed, 1959
,, ,,	*Lebistes reticulatus*	100	LC_{100}	48 hrs.	?	Schäperclaus, 1954
,, ,,	*Lebistes reticulatus*	10	LC_0	Indefinite	?	Schäperclaus, 1954
,, ,,	*Lepomis macrochirus*	11	LC_1	1 hr.	12°C	Wilford, 1967a
,, ,,	*Lepomis macrochirus*	30	LC_0	4 hrs.	12°C	Jackson, 1962
,, ,,	*Roccus saxatilis*	13	LC_{16}	48 hrs.	21°C	BSFW Fish Hatchery Report, 1968
,, ,,	*Roccus saxatilis*	13.5	LC_{15}	96 hrs.	21°C	Wellborn, 1971
,, ,,	*Salmo gairdneri*	13.3	LC_1	48 hrs.	12°C	Wilford, 1967a
,, ,,	*Salmo trutta*	22	LC_1	48 hrs.	12°C	Willford, 1967a
,, ,,	*Salvelinus fontinalis*	12	LC_1	48 hrs.	12°C	Willford, 1967a
,, ,,	*Salvelinus namaycush*	4.6	LC_1	48 hrs.	12°C	Willford, 1967a
Acrolein	*Lepomis macrochirus*	0.079	LC_{50}	24 hrs.	?	Johnson, 1968
,,	*Salmo trutta*	0.046	LC_{50}	24 hrs.	?	Johnson, 1968
Albucid	Unidentified	2,000	Toxic	96 hrs.	20°C	Schäperclaus, 1954
Albucid, Sodium	Unidentified	2,000	Not toxic	96 hrs.	20°C	Schäperclaus, 1954

[continued

Table 15. Known Toxicity Limits of Selected Chemicals used in Fish Parasite Control—*continued*

Compound	Fish Species	Dosage—ppm	Toxicity Limits	Exposure Time	Temperature	References
Ammonium nitrate	*Lepomis machrochirus*	800	LC_{100}	3.9 hrs.	?	McKee and Wolf, 1963
″ ″	*Lepomis macrochirus*	800	Not toxic	16 days	?	McKee and Wolf, 1963
″ ″	*Carassius auratus*	4,545	LC_{100}	90 hrs.	?	McKee and Wolf, 1963
Atabrine (= Atebrine)	*Ictalurus punctatus*	2	LC_0	1 hr.	25°C	Clemens and Sneed, 1959
″	*Ictalurus punctatus*	38	LC_1	48 hrs.	17°C	Willford, 1967a
″	*Ictalurus punctatus*	0.63	LC_0	72 hrs.	25°C	Clemens and Sneed, 1959
″	*Lebistes reticulatus*	10	Toxic	48 hrs.	12°C	Schäperclaus, 1954
″	*Lebistes reticulatus*	1	LC_0	48 hrs.	12°C	Schäperclaus, 1954
″	*Salmo gairdneri*	119	LC_1	48 hrs.	12°C	Willford, 1967a
″	*Salmo trutta*	130	LC_1	48 hrs.	12°C	Willford, 1967a
″	*Salvelinus fontinalis*	82	LC_1	48 hrs.	12°C	Willford, 1967a
″	*Salvelinus namaycush*	1.8	LC_1	48 hrs.	12°C	Willford, 1967a
″	Centrarchidae	0.2–0.8	Toxic	?	?	Berrios-Duran, *et al.*, 1964
″	*Lebistes reticulatus*	0.8	Toxic	?	?	Berrios-Duran, *et al.*, 1964
Bayer 9015	*Ictalurus punctatus*	1.0	Toxic	96 hrs.	?	R. Allison, 1969
Baygon	*Ictalurus punctatus*	1.0	Not toxic	96 hrs.	?	R. Allison, 1969
″	*Ictalurus punctatus*	5	Toxic	96 hrs.	?	R. Allison, 1969
Bayluscide	Many species	0.04–0.24	LC_{50}	?	?	Marking and Hogan, 1967
″	*Roccus saxatilis* (small)	0.78	LC_{16}	96 hrs.	21°C	Wellborn, 1971
″	*Tilapia* sp.	0.8	Toxic	?	?	Berrios-Duran, *et al.*, 1964
″	Trout and Minnows	0.22	Toxic	?	?	Reichenbach-Klinke, 1966

Baytex	Crayfish	0.23	LC_{100}	Indefinite	?	Osborn, 1966
,,	*Cyprinus carpio*	30	LC_{50}	?	?	Reichenbach-Klinke, 1966
,,	*Esox lucius*	0.6	LC_{50}	?	?	Reichenbach-Klinke, 1966
,,	*Ictalurus punctatus*	0.9	LC_{0}	?	?	Bishop, 1963
,,	*Lepomis macrochirus*	0.4–0.9	Toxic	?	?	Bishop, 1963
,,	Trout (young)	0.5	LC_{50}	?	?	Reichenbach-Klinke, 1966
Benzene hexachloride	*Cyprinus carpio*	0.02	LC_{0}	96 hrs.	24°C	F. Meyer, 1967
,, ,,	*Cyprinus carpio*	3.5	Toxic	....	....	Sarig and Lahav, 1958
,, ,,	*Ictalurus punctatus*	0.4	LC_{0}	?	10°C	Clemens and Sneed, 1959
,, ,,	*Ictalurus punctatus*	0.63	LC_{0}	?	20°C	Clemens and Sneed, 1959
,, ,,	*Ictiobus* sp.	0.02	LC_{0}	?	24°C	F. Meyer, 1967
,, ,,	*Notemigonus crysoleucas*	0.02	LC_{0}	96 hrs.	24°C	F. Meyer, 1967
,, ,,	*Notemigonus crysoleucas*	0.25	LC_{100}	96 hrs.	24°C	F. Meyer, 1967
,, ,,	Pond Fish	0.0125	Not toxic	?	?	Surber, 1948
,, ,,	*Salmo gairdneri*	0.0125	Toxic	?	?	Surber, 1948
,, ,,	*Salmo trutta*	0.0125	Not toxic	?	?	Surber, 1948
Bromex-50	*Carassius auratus*	2–5	LC_{50}	?	?	Ortho Co., 1969
,,	*Cyprinus carpio* (1–2 gm fish)	2	LC_{0}	?	?	Lahav, Sarig, and Shilo, 1966
,,	*Cyprinus carpio* (20–50 gm fish)	5	LC_{0}	?	?	Lahav, Sarig, and Shilo, 1966
,,	*Gambusia* sp.	0.19	Not toxic	Indefinite	?	Ortho Co., 1969
,,	*Lebistes reticulatus*	1–2	LC_{50}	?	?	Ortho Co., 1969
,,	*Lepomis macrochirus*	3.4	LC_{100}	?	?	Ortho Co., 1969
,,	*Micropterus salmoides*	3.4	LC_{100}	?	?	Ortho Co., 1969
,,	*Mugil cephalus*	0.1	Toxic	?	?	Lahav and Sarig, 1967
,,	*Salmo gairdneri*	0.08	LC_{50}	?	?	Ortho Co., 1969

[continued

Table 15. Known Toxicity Limits of Selected Chemicals used in Fish Parasite Control—*continued*

Compound	Fish Species	Dosage—ppm	Toxicity Limits	Exposure Time	Temperature	References
Cadmium sulfate	*Carassius auratus*	15	Not toxic	?	?	Deschiens, 1961
Calcium chloride	Fish	10,000	Toxic	?	?	Reichenbach-Klinke, 1966
Calcium hydroxide	Fish	100	Toxic	3–7 days	?	McKee and Wolf, 1963
Chevreul's salt	Fish	10	Not toxic	?	?	Gamet, *et al.*, 1964
,, ,,	*Lepomis auritus*	16	Toxic	....	....	Berrios-Duran, *et al.*, 1964
,, ,,	*Lebistes reticulatus*	2	LC_{25}	....	....	Berrios-Duran *et al.*, 1964
,, ,,	*Micropterus salmoides*	16	Not toxic	....	....	Berrios-Duran, *et al.*, 1964
,, ,,	*Tilapia* sp.	16	Not toxic	....	....	Berrios-Duran, *et al.*, 1964
Chloramine	*Cyprinus carpio* (young)	67	Slightly toxic	3 hrs.	?	Schäperclaus, 1954
,,	*Cyprinus carpio* (young)	10	Not toxic	Several days	?	Schäperclaus, 1954
Chloro-Phenoxytol (See Phenoxytol)	Fish	50	Toxic	?	?	Scott and Warren, 1964
Collargol	*Carassius vulgaris* (5 cm)	10	Toxic	24 hrs.	20°C	Schäperclaus, 1954
Copper sulfate (toxicity high in low carbonate waters; toxicity low in high carbonate waters)	*Carassius auratus*	0.5	LC_0	?	?	McKee and Wolf, 1963
,, ,,	*Cyprinus carpio*	0.33	LC_0	?	?	McKee and Wolf, 1963
,, ,,	*Cyprinus carpio*	0.33	Toxic	?	?	Reichenbach-Klinke, 1966
,, ,,	*Esox* sp.	0.4	LC_0	?	?	McKee and Wolf, 1963
,, ,,	Fish	0.5–160	LC_{100}	?	?	Surber, 1948
,, ,,	Fish	0.3	LC_0	Indefinite	22–24°C	Reichelt, 1971
,, ,,	*Ictalurus punctatus*	3.3	LC_1	24 hrs.	12°C	Reichelt, 1971
,, ,,	*Ictalurus punctatus*	0.4	LC_0	?	?	McKee and Wolf, 1963

” ”	*Lepomis macrochirus*	2.3	LC_1	48 hrs.	12°C	Reichelt, 1971
” ”	*Lepomis* sp.	1.35	LC_0	?	?	McKee and Wolf, 1963
” ”	*Lepomis* sp.	0.8	LC_0	?	?	McKee and Wolf, 1963
” ”	*Micropterus* sp.	0.8	LC_0	?	?	McKee and Wolf, 1963
” ”	*Micropterus* sp.	2.0	LC_0	?	?	McKee and Wolf, 1963
” ”	*Perca flavescens*	0.67	LC_0	?	?	McKee and Wolf, 1963
” ”	*Roccus saxatilis*	0.45	LC_{16}	96 hrs.	21°C	Wellborn, 1969
” ”	*Salmo trutta*	0.4	LC_1	48 hrs.	12°C	Reichelt, 1971
” ”	*Salmo* sp	3	LC_{100}	?	?	Kemp, 1958
” ”	*Salmo* sp.	0.14	Toxic	?	?	Reichenbach-Klinke, 1966
” ”	*Salvelinus fontinalis*	0.49	LC_1	48 hrs.	12°C	Reichelt, 1971
” ”	*Salvelinus namaycush*	0.1	LC_1	48 hrs.	12°C	Reichelt, 1971
” ”	Suckers	0.33	LC_0	?	?	McKee and Wolf, 1963
7-Co-ral	*Lepomis macrochrius*	0.18	LC_{50}	?	?	McKee and Wolf, 1963
”	*Pimephales promelas*	18	LC_{50}	96 hrs.	?	McKee and Wolf, 1963
”	*Roccus saxatilis* (small)	45	LC_{16}	96 hrs.	21°C	Wellborn, 1971
”	*Salmo gairdneri*	0.44	LC_1	48 hrs.	12°C	Willford, 1967a
”	*Ictalurus punctatus*	1–5	Not toxic	96 hrs.	?	R. Allison, 1969
Cuprous chloride	Fish	2	Not toxic	?	?	Deschiens and Floch, 1964
Cuprous oxide	Fish	5	Not toxic	?	?	Deschiens and Floch, 1964
DDT	*Cyprinus carpio* (fry)	0.057	LC_{50}	?	?	Reichenbach-Klinke, 1966
”	Fish	0.01–0.5	Toxic	?	?	Johnson, 1968
”	Fish	0.008–0.012	LC_{50}	?	?	Marking, 1966
”	*Ictalurus punctatus*	1	LC_0	1 hr.	19°C	Clemens and Sneed, 1959

[continued

Table 15. Known Toxicity Limits of Selected Chemicals used in Fish Parasite Control—*continued*

Compound	Fish Species	Dosage—ppm	Toxicity Limits	Exposure Time	Temperature	References
DDT	*Ictalurus punctatus*	0.18	LC_0	96 hrs.	19°C	Clemens and Sneed, 1959
″	*Micropterus* sp. (fry)	0.025	Toxic	?	?	Surber, 1948
Dipterex (See Dylox)						
Diquat	*Lepomis macrochrius*	11	Toxic	?	?	Wellborn, 1969
″	*Roccus saxatilis* (small)	70	LC_{16}	96 hrs.	21°C	Wellborn, 1969
DMSO	Fish	43,000	LC_{50}	?	?	Wilford, 1967b
Dursban	*Notemigonus crysoleucas*	0.02	LC_0	Indefinite	30°C	F. Meyer, Unpubl.
″	Fish	0.035	LC_{100}	....	....	Johnson, 1968
Dylox	*Anguilla japonica*	25.5	LC_{50}	96 hrs.	15–16°C	Kasahara, 1962
″	*Carassius auratus*	50	LC_{100}	2 hrs.	19–22°C	McKee and Wolf, 1963
″	*Carassius auratus*	40	LC_0	Indefinite	30°C	F. Meyer, 1966
″	*Carassius carassius*	27.0	LC_{50}	96 hrs.	16–19°C	Kasahara, 1962
″	*Cyprinus carpio*	27.5	LC_{50}	96 hrs.	16–20°C	Kasahara, 1962
″	*Cyprinus carpio* (young)	100	LC_{50}	?	?	Reichenbach-Klinke, 1966
″	*Esox* sp. (young)	1	LC_{50}	?	?	Reichenbach-Klinke, 1966
″	Fish	10	Not toxic	?	?	McKee and Wolf, 1963
″	*Notemigonus crysoleucas*	50	LC_0	Indefinite	30°C	F. Meyer, 1966
″	*Oryzias latipes*	79.2	LC_{50}	96 hrs.	16–20°C	Kasahara, 1962
″	*Pimephales promelas*	50	LC_0	Indefinite	30°C	F. Meyer, 1966
″	*Pimephales promelas*	180	LC_{100}	?	?	Johnson, 1968
″	*Pimephales promelas*	51	LC_{50} in hard water	?	25°C	McKee and Wolf, 1963
″	*Pimephales promelas*	180	LC_{50} in hard water	?	25°C	McKee and Wolf, 1963

″	*Pimephales promelas*	180	LC_{50}	96 hrs.	?	Johnson, 1968
″	*Roccus saxatilis* (young)	3.2	LC_{16}	96 hrs.	21°C	Wellborn, 1969
″	*Salmo gairdneri*	8.6	LC_{16}	48 hrs.	12°C	Willford, 1967a
″	*Salmo trutta*	6.7	LC_{1}	48 hrs.	12°C	Willford, 1967a
″	*Salmo* sp. (young)	0.8	LC_{50}	?	?	Reichenbach-Klinke, 1966
Eosin	*Rutilus* sp.; Trout	100	Not toxic	?	?	McKee and Wolf, 1963
Ethyl parathion (See Parathion)						
Ferbam	*Ictalurus punctatus*	20	LC_{0}	1 hr.	19°C	Clemens and Sneed, 1959
″	*Ictalurus punctatus*	0.8	LC_{0}	96 hrs.	19°C	Clemens and Sneed, 1959
Ferric sulfate	Fish	500	Toxic	?	?	Reichenbach-Klinke, 1966
Flagyl	*Ictalurus punctatus*	100	LC_{1}	48 hrs.	12–17°C	Willford, 1967a
″	*Lepomis macrochirus*	100	LC_{1}	48 hrs.	12–17°C	Willford, 1967a
″	*Salmo* sp.	100	LC_{1}	48 hrs.	12–17°C	Willford, 1967a
″	*Salvelinus* sp.	100	LC_{1}	48 hrs.	12–17°C	Willford, 1967a
Formalin	*Ictalurus punctatus*	316	LC_{0}	1 hr.	25°C	Clemens and Sneed, 1959
″	*Ictalurus punctatus*	126	LC_{0}	96 hrs.	25°C	Clemens and Sneed, 1959
″	*Ictalurus punctatus*	50	LC_{0}	96 hrs.	25°C	Leteux and Meyer, 1971
″	*Ictalurus punctatus*	167	LC_{50}	48 hrs.	?	Willford, 1967a
″	*Ictalurus punctatus*	126	LC_{100}	96 hrs.	25°C	Clemens and Sneed, 1959
″	*Lepomis macrochirus*	76	LC_{1}	48 hrs.	12°C	Willford, 1967a
	Oncorhynchus tschawytscha	28.2	LC_{100}	?	?	McKee and Wolf, 1963
″	*Roccus saxatilis*	12	LC_{16}	96 hrs.	21°C	Wellborn, 1969
″	*Roccus saxatilis*	50	LC_{0}	96 hrs.	25°C	Wellborn, 1969
″	*Salmo gairdneri*	121	LC_{1}	48 hrs.	12°C	Willford, 1967a

[*continued*

Table 15. Known Toxicity Limits of Selected Chemicals used in Fish Parasite Control—*continued*

Compound	Fish Species	Dosage—ppm	Toxicity Limits	Exposure Time	Temperature	References
Formalin	*Salmo trutta*	124	LC_1	48 hrs.	12°C	Wilford, 1967a
,,	*Salvelinus fontinalis*	124	LC_1	48 hrs.	12°C	Willford, 1967a
,,	*Salvelinus namaycush*	121	LC_1	48 hrs.	12°C	Willford, 1967a
Furanace	*Oncorhynchus* sp.	10	Toxic	?	?	Amend, 1971
Gesarol	Unidentified	20,000	Not toxic	$\frac{1}{2}$ hr.	?	Schäperclaus, 1954
Globucid	Unidentified	4,500	Not toxic	Several days	?	Schäperclaus, 1954
Globucid, sodium	Unidentified	4,500	Not toxic	Several days	?	Schäperclaus, 1954
Griseofulvin	Aquarium fishes	8	Not toxic	?	?	Anonymous, 1960
,,	Trout eggs	50	Toxic	?	?	Bradford, 1966
Hydrochloric acid	*Coregonus* sp.; *Esox* sp.	67	LC_0	?	?	Edminster and Gray, 1968
Hydrogen peroxide	Fish	25	LC_0	?	?	Reichenbach-Klinke, 1966
Lindane	*Cyprinus carpio* (young)	0.28	LC_{50}	?	?	Reichenbach-Klinke, 1966
,,	*Esox* sp. (young)	0.2	LC_{50}	?	?	Reichenbach-Klinke, 1966
,,	*Roccus saxatilis* (small)	0.32	LC_{16}	96 hrs.	21°C	Wellborn, 1971
,,	*Salmo* sp. (small)	0.3	LC_{50}	?	?	Reichenbach-Klinke, 1966
,,	*Salmo* sp.	0.38	LC_{50}	96 hrs.	20°C	Katz, 1961
Lysol	Trout	8	Toxic	?	?	Reichenbach-Klinke, 1966
Magnesium sulfate	*Anguilla* sp. (young)	12,000	LC_0	24 hrs.	?	McKee and Wolf, 1963
,, ,,	*Gambusia affinis*	15,500	Toxic	96 hrs.	....	McKee and Wolf, 1963
,, ,,	*Perca flavescens*	20–27,500	Toxic	72 hrs.	....	McKee and Wolf, 1963
Malachite green, oxalate	*Cyprinus carpio*	75	Not toxic	?	?	McKee and Wolf, 1963
,, ,, ,,	*Cyprinus carpio*	0.9	LC_0	?	?	Amlacher, 1961b

,, ,, ,,	*Ictalurus punctatus*	0.4	LC_0	1 hr.	25°C	Clemens and Sneed, 1959
,, ,, ,,	*Ictalurus punctatus*	0.1	LC_0	96 hrs.	25°C	Clemens and Sneed, 1959
,, ,, ,,	*Ictalurus punctatus*	0.075	LC_1	48 hrs.	17°C	Willford, 1967a
,, ,, ,,	*Lebistes reticulatus*	5	Toxic	?	?	McKee and Wolf, 1963
,, ,, ,,	*Lebistes reticulatus*	2	Toxic	?	?	Scott and Warren, 1964
,, ,, ,,	*Lepomis macrochirus*	0.064	LC_1	48 hrs.	12°C	Willford, 1967a
,, ,, ,,	*Lepomis macrochirus*	2	LC_0	1 hr.	12°C	Jackson, 1962
,, ,, ,,	*Oncorhynchus nerka* (small)	5	Not toxic	½ hr.	?	Rucker and Whipple, 1951
,, ,, ,,	*Oncorhynchus tschawytscha* (young)	1	Toxic	1hr.	?	J. Wood, 1968
,, ,, ,,	*Oncorhynchus tschawytscha* (adults)	1	Not toxic	1 hr.	?	J. Wood, 1968
,, ,, ,,	*Osphronemus* sp.	2	Toxic	?	?	Scott and Warren, 1964
,, ,, ,,	*Salmo gairdneri* (fry)	5	Not toxic	½ hr.	?	Rucker and Whipple, 1951
,, ,, ,,	*Salmo gairdneri*	0.22	LC_1	48 hrs.	12°C	Willford, 1967a
,, ,, ,,	*Salmo trutta*	0.25	LC_1	48 hrs.	12°C	Willford, 1967a
,, ,, ,,	*Salvelinus fontinalis*	0.16	LC_1	48 hrs.	12°C	Willford, 1967a
,, ,, ,,	*Salvelinus namaycush*	0.25	LC_1	48 hrs.	12°C	Willford, 1967a
,, ,, ,,	Trout	0.2	LC_0	?	?	Amlacher, 1961b
Malathion	*Cyprinus carpio* (young)	29.4	LC_{50}	?	?	Reichenbach-Klinke, 1966
,,	*Esox* sp. (young)	1	LC_{50}	....	....	Reichenbach-Klinke, 1966
,,	*Ictalurus punctatus*	400	LC_0	1 hr.	20°C	Clemens and Sneed, 1959
,,	*Ictalurus punctatus*	16	LC_0	96 hrs.	20°C	Clemens and Sneed, 1959
,,	*Ictiobus* sp.	0.13	Not toxic	?	?	Peterson, *et al.*, 1966
,,	*Lepisosteus* sp.	0.13	Not toxic	?	?	Peterson, *et al.*, 1966
,,	*Lepomis cyanellus*	0.5	Toxic	?	?	Bishop, 1963

[*continued*

Table 15. Known Toxicity Limits of Selected Chemicals used in Fish Parasite Control—*continued*

Compound	Fish Species	Dosage—ppm	Toxicity Limits	Exposure Time	Temperature	References
Malathion	*Lepomis macrochirus*	0.13	Toxic	?	?	Peterson, *et al.*, 1966
,,	*Lepomis macrochirus*	0.2	Toxic	?	?	Mount and Stephan, 1967
,,	*Micropterus salmoides*	0.13	Toxic	?	?	Peterson, *et al.*, 1966
,,	*Micropterus salmoides*	0.5	Toxic	?	?	Bishop, 1963
,,	*Pimephales promelas*	12.5	LC_{50}	96 hrs.	20°C	Johnson, 1968
,,	*Pimephales promelas*	0.2	Not toxic	?	?	Mount and Stephan, 1967
,,	*Pylodictis olivaris*	0.5	Not toxic	?	?	Bishop, 1963
,,	*Pylodictis olivaris*	0.13	Not toxic	?	?	Peterson, *et al.*, 1966
,,	*Roccus chrysops*	0.13	Toxic	?	?	Peterson, *et al.*, 1966
,,	*Roccus saxatilis* (small)	0.16	LC_{15}	96 hrs.	21°C	Wellborn, 1971
,,	*Salmo* sp. (young)	1	LC_{50}	?	?	Reichenbach-Klinke, 1966
,,	*Salvelinus fontinalis*	1	Toxic	?	?	Larsen, 1963
,,	*Tilapia* sp.	0.5	Not toxic	?	?	Bishop, 1963
Mercuric chloride	*Gasterosteus aculeatus*	0.008	Toxic	?	?	Reichenbach-Klinke, 1966
,, ,,	Fish	0.05	Toxic	?	?	McKee and Wolf, 1963
Methylene blue	*Ictalurus punctatus*	68.5	LC_1	48 hrs.	17°C	Johnson, 1968
,, ,,	*Lepomis macrochirus*	13.5	LC_1	48 hrs.	12°C	Johnson, 1968
,, ,,	*Salmo gairdneri*	10.7	LC_1	48 hrs.	12°C	Johnson, 1968
,, ,,	*Salmo trutta*	17.8	LC_1	48 hrs.	12°C	Johnson, 1968
,, ,,	*Salvelinus fontinalis*	11.1	LC_1	48 hrs.	12°C	Johnson, 1968
,, ,,	*Salvelinus namaycush*	22.7	LC_1	48 hrs.	12°C	Johnson, 1968

Micropur 1000	*Carassius vulgaris*	10	Toxic to weak fish	?	?	Schäperclaus, 1954
,, ,,	*Carassius vulgaris*	6.7	Not toxic	?	?	Schäperclaus, 1954
Nickel sulfate	*Ictalurus punctatus*	66	LC_1	48 hrs.	17°C	Willford, 1967a
,, ,,	*Lepomis macrochirus*	300	LC_1	48 hrs.	12°C	Willford, 1967a
,, ,,	*Salmo gairdneri*	125	LC_1	48 hrs.	12°C	Willford, 1967a
,, ,,	*Salmo trutta*	138	LC_1	48 hrs.	12°C	Wilford, 1967a
,, ,,	*Salvelinus fontinalis*	174	LC_1	48 hrs.	12°C	Willford, 1967a
,, ,,	*Salvelinus namaycush*	24.6	LC_1	48 hrs.	12°C	Willford, 1967a
Nicotine sulfate	Fish	3	Toxic	?	?	Reichenbach-Klinke, 1966
Oxytetracycline (See Terramycin)						
Paraquat	Pond fish	1.14	Toxic	1–16 days	20–25°C	Earnest, 1971
Parathion (Ethyl)	*Cyprinus carpio* (young)	3.5	LC_{50}	?	?	Reichenbach-Klinke, 1966
,,	*Esox* sp. (young)	3.0	LC_{50}	?	?	Reichenbach-Klinke, 1966
,,	Fish	0.25	Not toxic	?	?	Osborn, 1966
,,	*Lepomis macrochirus*	0.25	Toxic	?	?	Osborn, 1966
,,	*Lepomis macrochirus*	0.063	Toxic	?	?	Surber, 1948
,,	*Micropterus salmoides*	0.25	Toxic	?	?	Osborn, 1966
,,	*Pimephales promelas*	2.7	LC_{100}	?	?	Johnson, 1968
,,	*Salmo gairdneri*	0.38	Not toxic	?	?	Surber, 1948
Phenoxetol	*Lebistes reticulatus*	100	LC_0	?	?	Scott and Warren, 1964
,,	*Osphronemus* sp.	100	LC_0	?	?	Scott and Warren, 1964
Plasmochin	*Lebistes reticulatus*	1	Toxic	72 hrs.	20°C	Schäperclaus, 1954

[continued

Table 15. Known Toxicity Limits of Selected Chemicals used in Fish Parasite Control—*continued*

Compound	Fish Species	Dosage—ppm	Toxicity Limits	Exposure Time	Temperature	References
PMA	*Ictalurus punctatus*	37.6	LC_0	1 hr.	24°C	Clemens and Sneed, 1959
,,	*Ictalurus punctatus*	0.37	LC_0	72 hrs.	24°C	Clemens and Sneed, 1959
,,	*Ictalurus punctatus*	1.95	LC_1	48 hrs.	17°C	Willford, 1967a
,,	*Lepomis macrochirus*	8.4	LC_1	48 hrs.	17°C	Willford, 1967a
,,	*Oncorhynchus nerka*	10	Not toxic	1 hr.	?	Rucker and Whipple, 1951
,,	*Salmo gairdneri*	1.4	LC_1	48 hrs.	12°C	Willford, 1967a
,,	*Salmo gairdneri* (fry)	10	Not toxic	1 hr.	?	Rucker and Whipple, 1951
,,	*Salmo gairdneri* (young)	5	Toxic	1 hr.	?	Rucker and Whipple, 1951
,,	*Salmo trutta*	5.1	LC_1	48 hrs.	12°C	Willford, 1967a
,,	*Salvelinus fontinalis*	7.8	LC_1	48 hrs.	12°C	Willford, 1967a
,,	*Salvelinus namaycush*	4.5	LC_1	48 hrs.	12°C	Willford, 1967a
Potassium permanganate	*Anguilla* sp. (young)	11.8	Toxic	8 hrs.	?	McKee and Wolf, 1963
,, ,,	*Carassius auratus*	10	Toxic	18 hrs.	?	McKee and Wolf, 1968
,, ,,	*Carassius auratus*	6	LC_{100}	?	?	Peterson, *et al.*, 1966
,, ,,	*Dorosoma cepedianum*	3	Not toxic	?	?	Peterson, *et al.*, 1966
,, ,,	Fish	3	LC_0	?	?	Reichenbach-Klinke, 1966
,, ,,	*Gambusia affinis*	12	Toxic	24 hrs.	?	McKee and Wolf, 1963
,, ,,	*Ictalurus punctatus*	9.1	LC_0	1 hr.	25°C	Clemens and Sneed, 1959
,, ,,	*Ictalurus punctatus*	3.2	LC_0	24 hrs.	25°C	Clemens and Sneed, 1959
,, ,,	*Lebistes reticulatus*	20	Toxic	?	?	Scott and Warren, 1964
,, ,,	*Lepomis macrochirus*	3	LC_{100}	?	?	Lawrence, 1956
,, ,,	*Lepomis macrochirus*	3	Toxic	?	?	McKee and Wolf, 1963

″ ″	*Lepomis macrochirus*	5.2	Toxic	24 hrs.	?	McKee and Wolf, 1963
″ ″	*Micropterus salmoides*	4.0	Toxic	?	?	McKee and Wolf, 1963
″ ″	*Micropterus salmoides*	4.0	LC_{100}	?	?	Peterson, *et al.*, 1966
″ ″	*Osphronemus* sp.	20	Toxic	?	?	Scott and Warren, 1964
″ ″	*Pimephales promelas*	5	Toxic	?	?	McKee and Wolf, 1963; Peterson, *et al.*, 1966
″ ″	*Roccus saxatilis* (young)	1.7	LC_{16}	96 hrs.	21°C	Wellborn, 1969
″ ″	*Stizostedion* sp.	3	Toxic	?	?	Peterson, *et al.*, 1966
″ ″	Trout	6.25	Toxic	24 hrs.	?	McKee and Wolf, 1963
Quinine hydrochloride	*Cyprinus carpio* (fry)	10	Toxic	24 hrs.	20°C	Schäperclaus, 1954
″ ″	*Ictalurus punctatus*	100	LC_1	48 hrs.	17°C	Willford, 1967a
″ ″	*Lepomis macrochirus*	100	LC_1	48 hrs.	17°C	Willford, 1967a
″ ″	*Salmo gairdneri*	100	LC_1	48 hrs.	17°C	Willford, 1967a
″ ″	*Salmo trutta*	100	LC_1	48 hrs.	17°C	Willford, 1967a
″ ″	*Salvelinus fontinalis*	100	LC_1	48 hrs.	17°C	Willford, 1967a
″ ″	*Salvelinus namaycush*	100	LC_1	48 hrs.	17°C	Willford, 1967a
Quinine sulfate	*Ictalurus punctatus*	100	LC_0	1 hr.	23°C	Clemens and Sneed, 1959
″ ″	*Ictalurus punctatus*	23.7	LC_0	96 hrs.	23°C	Clemens and Sneed, 1959
Rivanol	*Ictalurus punctatus*	5.6	LC_0	1 hr.	20°C	Clemens and Sneed, 1959
″	*Ictalurus punctatus*	1.8	LC_0	96 hrs.	20°C	Clemens and Sneed, 1959
″	*Lebistes reticulatus*	10	Toxic	?	20°C	Schäperclaus, 1954
Roccal	*Carassius auratus*	2	Not toxic	7 days	?	Kincheloe, 1964
″	*Ictalurus punctatus*	0.74	LC_1	48 hrs.	17°C	Willford, 1967a
″	*Lepomis macrochirus*	1.21	LC_1	48 hrs.	12°C	Willford, 1967a
″	*Salmo gairdneri*	1.32	LC_1	48 hrs.	12°C	Willford, 1967a

[continued

Table 15. Known Toxicity Limits of Selected Chemicals used in Fish Parasite Control—*continued*

Compound	Fish Species	Dosage—ppm	Toxicity Limits	Exposure Time	Temperature	References
Roccal	*Salmo gairdneri*	2	Not toxic	1 hr.	?	Rucker and Whipple, 1951
"	*Salmo trutta*	1.05	LC_1	48 hrs.	12°C	Willford, 1967a
"	*Salvelinus fontinalis*	2.05	LC_1	48 hrs.	12°C	Willford, 1967a
"	*Salvelinus namaycush*	1.1	LC_1	48 hrs.	12°C	Willford, 1967a
Ronnel	*Ictalurus punctatus*	0.74	LC_1	48 hrs.	17°C	Willford, 1967a
"	*Lepomis macrochirus*	0.133	LC_1	48 hrs.	12°C	Willford, 1967a
"	*Salmo gairdneri*	0.43	LC_1	48 hrs.	12°C	Willford, 1967a
"	*Salmo trutta*	0.144	LC_1	48 hrs.	12°C	Willford, 1967a
"	*Salvelinus fontinalis*	0.12	LC_1	48 hrs.	12°C	Willford, 1967a
"	*Salvelinus namaycush*	0.31	LC_1	48 hrs.	12°C	Willford, 1967a
Ruelene	*Ictalurus punctatus*	26.9	LC_1	48 hrs.	17°C	Willford, 1967a
"	*Lepomis macrochirus*	26.4	LC_1	48 hrs.	12°C	Willford, 1967a
"	*Salmo gairdneri*	28.4	LC_1	48 hrs.	12°C	Willford, 1967a
"	*Salmo trutta*	21.7	LC_1	48 hrs.	12°C	Wilford, 1967a
"	*Salvelinus fontinalis*	26.6	LC_1	48 hrs.	12°C	Willford, 1967a
"	*Salvelinus namaycush*	17.0	LC_1	48 hrs.	12°C	Willford, 1967a
"	*Ictalurus punctatus*	1.0	Not toxic	96 hrs.	?	R. Allison, 1969
"	*Ictalurus punctatus*	5.0	Toxic	96 hrs.	?	R. Allison, 1969
Silver nitrate	*Gasterosteus aculeatus*	0.004	Toxic	?	?	Reichenbach-Klinke, 1966
Sodium chloride	*Anguilla* sp. (young)	11,000	Not toxic	48 hrs.	?	McKee and Wolf, 1963
" "	*Carassius auratus*	5,000	Not toxic	Indefinite	?	McKee and Wolf, 1963
" "	*Coregonus* sp. (fry)	17,000	Not toxic	?	?	Edminster and Gray, 1948
" "	Cyprinids	14,000	Not toxic	72 hrs.	?	Nakai and Kokai, 1931

,, ,,	Fish	15,000	Not toxic	?	?	Roth, in Mellen, 1928
,, ,,	*Gasterosteus* sp.	20,000	Not toxic	?	?	McKee and Wolf, 1963
,, ,,	*Ictalurus catus, Ictalurus furcatus, Ictalurus punctatus*	11,000	Not toxic	Indefinite	?	Allen and Avault, 1970
,, ,,	*Lepomis macrochirus*	4,000	Not toxic	?	?	Tebo and McCoy, 1964
,, ,,	*Micropterus salmoides*	4,000	Not toxic	?	?	Tebo and McCoy, 1964
,, ,,	*Oncorhynchus nerka* (young)	30,000	Not toxic	?	?	Rucker and Whipple, 1951
,, ,,	*Oncorhynchus* sp. (eggs and fry)	5,850	Not toxic	?	?	McKee and Wolf, 1963
,, ,,	*Perca flavescens*	17,500	Not toxic	Indefinite	?	McKee and Wolf, 1963
,, ,,	*Perca fluviatilis*	10,000	Not toxic	Indefinite	?	Herbert and Mann, 1958
,, ,,	*Rutilus rutilus*	15,000	Not toxic	?	?	Herbert and Mann, 1958
,, ,,	*Salmo gairdneri* (fry)	30,000	Toxic	?	?	Rucker and Whipple, 1951
,, ,,	*Salmo gairdneri* (young)	30,000	Not toxic	?	?	Rucker and Whipple, 1951
,, ,,	*Salvelinus fontinalis*	30,000	Not toxic	1 hr.	?	McKee and Wolf, 1963
,, ,,	*Stizostedion vitreum*	4,000	Not toxic	?	?	Edminster and Gray, 1948
,, ,,	*Tilapia* sp.	53,500	Not toxic	?	?	Lotan, 1960
,, ,,	Trout	20,000	Not toxic	Indefinite	?	L. Allison, 1950
,, ,,	Trout (adults)	30,000	Not toxic	?	?	McKee and Wolf, 1963
Sodium chlorite	Freshwater fish	12	Not toxic	?	?	Garibaldi, 1971
Sodium pentachlorophenate	*Cyprinus carpio*	8	Toxic	70 min.	?	Reichenbach-Klinke, 1966
,, ,,	*Ictalurus punctatus*	2.5	LC_0	1 hr.	25°C	Clemens and Sneed, 1959
,, ,,	*Ictalurus punctatus*	0.25	LC_0	96 hrs.	25°C	Clemens and Sneed, 1959
,, ,,	*Pimephales promelas*	0.25	Not toxic	?	?	Crandall and Goodnight, 1959
,, ,,	*Pimephales promelas*	0.4	Toxic	?	?	Crandall and Goodnight, 1959
,, ,,	Trout	8	Toxic	20 min.	?	Reichenbach-Klinke, 1966

[continued

Table 15. Known Toxicity Limits of Selected Chemicals used in Fish Parasite Control—*continued*

Compound	Fish Species	Dosage—ppm	Toxicity Limits	Exposure Time	Temperature	References
Sulfamethazine	*Ictalurus punctatus*	100	LC_1	48 hrs.	17°C	Willford, 1967a
,,	*Lepomis macrochirus*	100	LC_1	48 hrs.	12°C	Willford, 1967a
,,	*Salmo gairdneri*	100	LC_1	48 hrs.	12°C	Willford, 1967a
,,	*Salmo trutta*	100	LC_1	48 hrs.	12°C	Willford, 1967a
,,	*Salvelinus fontinalis*	100	LC_1	48 hrs.	12°C	Willford, 1967a
,,	*Salvelinus namaycush*	100	LC_1	48 hrs.	12°C	Willford, 1967a
Sulfuric acid	*Coregonus* sp. (fry)	80	Not toxic	?	?	Edminster and Gray, 1948
,, ,,	*Stizostedion vitreum* (fry)	71	Not toxic	?	?	Edminster and Gray, 1948
Tag	*Ictalurus punctatus*	4	LC_0	1 hr	20°C	Clemens and Sneed, 1959
,,	*Ictalurus punctatus*	0.4	LC_0	96 hrs.	20°C	Clemens and Sneed, 1959
Terramycin	*Roccus saxatilis* (small)	125	LC_{16}	96 hrs.	21°C	Wellborn, 1969 and 1971
TFM	*Carassius auratus*	10	LC_0	96 hrs.	28°C	F. Meyer, Unpubl.
,,	*Carassius auratus*	18	LC_{100}	96 hrs.	28°C	F. Meyer, Unpubl.
,,	*Catostomus commersoni*	5	LC_{25}	24 hrs.	13°C	Applegate and King, 1962
,,	*Ictalurus natalis*	5.75	LC_{25}	24 hrs.	13°C	Applegate and King, 1962
,,	*Ictalurus punctatus*	10	LC_0	96 hrs.	28°C	F. Meyer, Unpubl.
,,	*Ictalurus punctatus*	18	LC_{100}	96 hrs.	28°C	F. Meyer, Unpubl.
,,	*Lepomis macrochirus*	21.5	LC_{25}	24 hrs.	13°C	Applegate and King, 1962
,,	*Micropterus dolomieu*	34.5	LC_{25}	24 hrs.	13°C	Applegate and King, 1962
,,	*Micropterus salmoides*	22.0	LC_{25}	24 hrs.	13°C	Applegate and King, 1962
,,	*Notemigonus crysoleucas*	14.75	LC_{25}	24 hrs.	13°C	Applegate and King, 1962
,,	*Notropis heterolepis*	13.25	LC_{25}	24 hrs.	13°C	Applegate and King, 1962

,,	*Perca flavescens*	7.25	LC_{25}	24 hrs.	13°C	Applegate and King, 1962
,,	*Pimephales promelas*	16.0	LC_{25}	24 hrs.	13°C	Applegate and King, 1962
,,	*Rana catesbeiana* (tadpoles)	8.0	LC_{0}	96 hrs.	28°C	F. Meyer, Unpubl.
,,	*Rana catesbeiana* (tadpoles)	14.0	LC_{100}	96 hrs.	28°C	F. Meyer, Unpubl.
,,	*Salmo gairdneri*	12.0	LC_{25}	24 hrs.	13°C	Applegate and King, 1962
,,	*Stizostedion vitreum*	5.75	LC_{25}	24 hrs.	13°C	Applegate and King, 1962
Thiram	*Ictalurus punctatus*	1	LC_{0}	1 hr.	19°C	Clemens and Sneed, 1959
,,	*Ictalurus punctatus*	0.63	LC_{0}	96 hrs.	19°C	Clemens and Sneed, 1959
Tiguvon (See also Baytex)	*Ictalurus punctatus*	2.9	LC_{1}	48 hrs.	17°C	Willford, 1967a
,,	*Lepomis macrochirus*	4.1	LC_{1}	48 hrs.	12°C	Willford, 1967a
,,	*Salmo gairdneri*	2.2	LC_{1}	48 hrs.	12°C	Willford, 1967a
,,	*Salmo trutta*	1.8	LC_{1}	48 hrs.	12°C	Willford, 1967a
,,	*Salvelinus fontinalis*	4.3	LC_{1}	48 hrs.	12°C	Willford, 1967a
,,	*Salvelinus namaycush*	4.1	LC_{1}	48 hrs.	12°C	Willford, 1967a
,,	*Ictalurus punctatus*	1.0	Not toxic	96 hrs.	?	R. Allison, 1969
,,	*Ictalurus punctatus*	5.0	Toxic	96 hrs.	?	R. Allison, 1969
n-Tritylmorpholine	*Barbus* sp.	0.1	LC_{100}	?	?	Boyce, *et al.*, 1967
,,	*Tilapia* sp.	0.1	LC_{100}	?	?	Boyce, *et al.*, 1967
Trypaflavine (acriflavine hydrochloride)	*Lebistes reticulatus*	100	Toxic	48 hrs.	20°C	Schäperclaus, 1954
,,	*Lebistes reticulatus*	10	Not toxic	96 hrs.	20°C	Schäperclaus, 1954
TV-1096	*Ictalurus punctatus*	17.8	LC_{1}	48 hrs.	17°C	Willford, 1967a
,,	*Lepomis macrochirus*	18.2	LC_{1}	48 hrs.	12°C	Willford, 1967a
,,	*Salmo gairdneri*	12	LC_{1}	48 hrs.	12°C	Willford, 1967a
,,	*Salmo trutta*	10	LC_{1}	48 hrs.	12°C	Willford, 1967a

[*continued*

Table 15. Known Toxicity Limits of Selected Chemicals used in Fish Parasite Control—*continued*

Compound	Fish Species	Dosage—ppm	Toxicity Limits	Exposure Time	Temperature	References
TV-1096	*Salvelinus fontinalis*	12.4	LC_1	48 hrs.	12°C	Willford, 1967a
,,	*Salvelinus namaycush*	6.7	LC_1	48 hrs.	12°C	Willford, 1967a
Zectran	*Notemigonus crysoleucas*	10.0	LC_{100}	96 hrs.	28°C	F. Meyer, Unpubl
,,	*Notemigonus crysoleucas*	1.0	LC_0	96 hrs.	28°C	F. Meyer, Unpubl.
Ziram	*Ictalurus punctatus*	2.0	LC_0	1 hr.	19°C	Clemens and Sneed, 1959
,,	*Ictalurus punctatus*	0.2	LC_0	96 hrs.	19°C	Clemens and Sneed, 1959

Table 16. CHEMICAL NAMES AND SYNONYMS

Name	Structure or Chemical Name	Synonyms	Common Use
ACETARSONE	n-Acetyl-4-hydroxy-n-arsanilic acid	Acetarsol Stovarsol Stovarsolan	Protozoicide
ACETIC ACID, commercial grade	CH_3COOH	....	Industrial chemical
ACRIFLAVINE, hydrochloride	Acid salt of Acriflavine	Trypaflavin	Dye, bacteriostat
ACRIFLAVINE, neutral	Mixture of 2,8-diamino-10-methylacridinium chloride and 2,8 diaminoacridine	....	Dye, bacteriostat
ACROLEIN	2-Propenal	Acraldehyde Aqualin	Herbicide
ALBUCID	N-acetylsulfanilamide	Sulfacetamide Sulfacyl	Antimicrobial
ALUMINUM SULFATE, commercial grade	$Al_2(SO_4)_3 \cdot nH_2O$	Alum	Industrial chemical
AMMINOSIDIN	....	Component of Emtrysidina	Antibiotic
AMMONIA, 28% in water	NH_3 or NH_4OH	Ammonium hydroxide	Industrial chemical
AMMONIUM CARBONATE	A mixture of ammonium bicarbonate and ammonium carbamate	Hartshorn	Industrial chemical
AMMONIUM CHLORIDE, commercial grade	NH_4Cl	Darammon	Industrial chemical
AMMONIUM HYDROXIDE	NH_4OH	....	Industrial chemical
AMMONIUM NITRATE, commercial grade	NH_4NO_3	....	Industrial chemical
AMOPYROQUIN, Parke, Davis & Co.	4-(7-chloro-4-quinolylamino)-1-pyrrolidyl-o-cresol dihydrochloride	Propoquin	Protozoicide
AMPROLIUM	1-[(4-Amino-2 propyl-5-pyrimidinyl) methyl]-2-picolinium chloride hydrochloride	Amprol	Protozoicide
ANTHIUM DIOXCIDE	5% chlorine dioxide	Microcide	Unknown
ANTIMYCIN A	Antibiotic derived from *Streptomyces* sp.	Fintrol	Fish toxicant

[continued

Table 16. Chemical Names and Synonyms—*continued*

Name	Structure or Chemical Name	Synonyms	Common Use
AQUA-AID	Ingredients are listed	....	Aquarist trade product
AQUAROL	Unknown commercial formulation	....	Unknown
ATABRINE, hydrochloride	3-Chloro-7-methyoxy-9-(1-methyl-4-diethyl-aminobutyl-amino) acridine dihydrochloride	Atebrine Chinacrine Mepacrine Quinacrine	Malariastat
AUREOMYCIN	7-chloro-4-dimethylamino-1,4,4a,5,5a,6,11,12a-octahydro-3,6,10,12, 12a-pentahydroxy-6-methyl-1,11, dioxo-2-naphtha-cenecarboxamide	Biomitsin Biomycin Chlortetracycline	Antibiotic
BALSAM OF PERU OIL	Benzyl cinnamate—extract from *Myroxylon pereirae*	Balsam of Peru China oil Peru oil	Insecticide, Industrial chemical
BARIUM CARBONATE	$BaCO_3$	....	Industrial chemical
BARIUM CHLORIDE	$BaCl_2 \cdot 2H_2O$	....	Industrial chemical
BASIC BRIGHT GREEN, oxalate	Tetra-ethyl-diamino-triphenyl carbinol	Brilliant green	Dye
BASIC VIOLET	Unknown	....	Russian dye (Musselius and Filippova, 1968)
®BAYER 73	See Bayluscide	Bayluscide	Molluscicide
®BAYGON, 50% wettable powder	o-Isopropoxyphenyl methyl-carbamate	Propoxur Uden	Insecticide
®BAYLUSCIDE	Aminoethanol dichloronitro-salicylanilide (ethanolamine salt of niclosamide)	Bayer 73 Bayluscid®	Molluscicide
®BAYTEX	O,o-dimethyl O-[4-(methylthio)-m-tolyl] phosphorothioate	Bayer 29493 Entex Fenthion Tiguvon	Insecticide

BENZENE HEXACHLORIDE	1,2,3,4,5,6-hexachlorocyclohexane	BHC Gammexane Lindane Lexone 666 Tri-6	Insecticide
BETADINE	Iodine + polyvinylpyrrolidone complex	Povidone-Iodine PVP-I	Disinfectant
BETANAPHTHOL	B-hydroxynaphthalene	Isonaphthol 2-naphthol	Antiseptic and parasiticide
BIOCIDAL RUBBER	See butyl tin oxide and di-N-butyl tin oxide	....	Barnacle control
BIOMYCIN	See Aureomycin	....	Antibiotic
BIPYRIDILIUM	See Diquat and Paraquat	....	Herbicide
BIS-OXIDE	Tri-n-butyl tin oxide	....	Anthelmintic
BITHIONOL	2',2-Thiobis (4,6-dichlorophenol)	Actamer Bithin Lorothidol TBP	Antimicrobial
BORIC ACID	H_3BO_3	....	Disinfectant
BRIGHT GREEN	See malachite green	....	Dye
BROMEX-50	1,2 dibromo-2,-dichloro-ethyl dimethyl phosphate	Dibrom Naled	Insecticide
BUTYL TIN OXIDE	See di-n-butyl tin oxide	....	Anthelmintic
BUTYNORATE	See dibutyltin dilaurate	Tinostat	Anthelmintic
CADMIUM SULFATE	$CdSO_4 \cdot n\ H_2O$	....	Industrial chemical
CALCIUM CARBONATE, commercial grade	$CaCO_3$ (usually magnesium carbonate—$MgCO_3$)	Lime	Industrial chemical
CALCIUM CYANAMIDE	NCNCa	Cyanamide	Agricultural chemical

[continued

Table 16. Chemical Names and Synonyms—*continued*

Name	Structure or Chemical Name	Synonyms	Common Use
CALCIUM HYDROXIDE, commercial grade	$Ca(OH)_2$	Hydrated lime Slaked lime	Industrial chemical
CALCIUM HYPOCHLORITE	$Ca(OCl)_2$	Chlorinated lime HTH Perchloron	Disinfectant
CALCIUM OXIDE	CaO	Quicklime Unslaked lime	Industrial chemical
CALOMEL	Mercurous chloride	Calomelol	Fungicide
CARBARSONE	N-carbamoylarsanilic acid	Arthinol	Protozoicide
CARBARSONE OXIDE	p-Carbamidophenyl arsenoxide	....	Protozoicide
CASTOR BEAN PLANT	*Ricinus communis*	....	Unknown
CHELATED COPPER	$CuSO_4$ + EDTA	....	Algicide
CHEVREUL'S SALT	$Cu_2SO_3 \cdot CuSO_3 \cdot 2H_2O$	....	Molluscicide
CHLORAMINE	See Chloramine-T	Chloramin	Antiseptic
CHLORAMINE-B	Sodium benzenesulfonchloramide	....	Disinfectant
CHLORAMINE-T	Sodium p-toluenesulfonchloramide	Chloramin Chloramine Chlorazene Chlorozone Halamid	Antiseptic
CHLORAMPHENICOL	D(-)-threo-2,2-Dichloro-N-[B-hydroxy-a-(hydroxymethyl)-p-nitrophenethyl] acetamide	Chloromycetin	Antibiotic
CHLORINE	As Cl_2 gas or available element	....	Disinfectant
CHLOROPHOS	See Dylox	....	Insecticide
CHLORTETRACYCLINE	See Aureomycin	....	Antibiotic

CITRIC ACID	2-Hydroxy-1,2,3-propanetricarboxylic acid	....	Industrial chemical
COLLARGOL	Colloidal silver	Argentum Crede'	Disinfectant
COPPER CARBONATE	$CuCO_3 \cdot Cu(OH)_2 \cdot H_2O$	Cupric carbonate	Seed treatment
COPPER, micronized	....	....	Industrial chemical
COPPER SULFATE	$CuSO_4 \cdot 5\ H_2O$	Bluestone	Algicide
COPPER SULFATE plus ACETIC ACID	$CuSO_4 + CH_3COOH$	....	Algicide
7-CO-RAL	0,0-Diethyl 0-(3-chloro-4-methyl-2-oxo-2H-1-benzopyran-7-yl) phosphorothioate	Co-Ral Coumaphos Muscatox Resitox	Insecticide
COTTONSEED OIL	*Gossypium* sp.	....	Feedstuff
CUPRIC ACETATE	$Cu(CH_3COO)_2$	Verdigris	Industrial chemical
CUPRIC CHLORIDE	$CuCl_2$	....	Industrial chemical
CUPRIC SULFATE, anhydrous	See copper sulfate	....	Industrial chemical
CUPROUS CHLORIDE	CuCl	Nantokite	Molluscicide
CUPROUS OXIDE	Cu_2O	....	Industrial chemical
CUTRINE	Copper sulfate + triethanolamine and other additives (commercial formulation)	....	Algicide
CYPREX	n-Dodecylquanidine acetate	Dodine	Fungicide
CYZINE	2-Acetylamino-5-nitrothiazole	Trichorad	Protozoicide
DARAPRIM	2,4-Diamino-5-(p-chlorophenyl)-6-ethyl-primidine	Pyrimethamine	Antimalarial
DDFT	Diflurodiphenyltrichloroethane	DFDT	Insecticide
DDT	1,1,1-Trichloro-2,2-bis(p-chlorophenyl) ethane	Gesarol	Insecticide
DDVP	0,0-dimethyl 0-(2,2-dichlorovinyl) phosphate	Dichlorovos Vapona	Insecticide
DFDT	Difluorodiphenyltrichloroethane	DDFT	Insecticide

[continued

Table 16. Chemical Names and Synonyms—*continued*

Name	Structure or Chemical Name	Synonyms	Common Use
DIBROM, emulsifiable concentrate	See Naled	Bromex-50 Naled	Insecticide
DIBUTYLTIN DILAURATE	$(C_4H_9)_2Sn\ [O_2C(CH_2)_{10}\ CH_3]_2$	Butynorate Tinostat	Anthelmintic
DIESEL FUEL, commercial grade	....	....	Transportation fuel
DIMETRIDAZOLE	1,2-Dimethyl-5-nitroimidazole	Emtryl	Protozoicide
DI-N-BUTYL TIN OXIDE	di-n-butyl-oxo-stannate	....	Anthelmintic
DIPTEREX, 5% wettable powder	See Dylox	....	Insecticide
DIQUAT	1,1'-Ethylene-2,2'-dipyridylium dibromide	Reglone	Herbicide
DI-SYSTON	0,0-diethyl S-[2-(ethylthio) ethyl] phosphorothioate	Demeton-S Isosystox	Insecticide
DMSO	$(CH_3)_2\ SO$	Dimethyl Sulfoxide	Penetrant and carrier
DOWCO 212	Experimental compound (Dow Chemical Co.)	....	Molluscicide
DOWICIDE G	See Sodium pentachlorophenate	....	Wood preservative
DURSBAN	0,0, Diethyl-0-3,5,6-trichloro-2-pyridyl phosphorothioate	....	Insecticide
DVP	Believed to be DDVP	....	Insecticide
DYLOX	0,0-Dimethyl 2,2,2-trichloro-1-hydroxyethyl phosphonate	Chlorofos Chlorophos Dipterex Foschlor Masoten Metrifonat Neguvon Trichlorofon	Insecticide

EMTRYSIDINA	N'-(3-Methoxy-2-purazinyl) sulfanilamide plus amminosidin	Kelfizina Sulfalene	Antimicrobial
ENHEPTIN	2-amino-5-nitrothiazol	....	Protozoicide
ENTOBEX	4,7-Phenanthrolin-5,6-dione	Phanquone	Protozoicide
EOSIN	2',4',5',7'-Tetrabromofluorescein	....	Dye
ETHYLENE GLYCOL	1,2 Ethanediol	....	Antifreeze
ETHYL PARATHION	0,0-Diethyl 0-p-nitrophenyl phosphorothioate	Niran Parathion Phoskil Thiophos	Insecticide
FERBAM	Ferric dimethyldithiocarbamate	Fermate	Fungicide
FERRIC SULFATE	$Fe_2(SO_4)_3$	....	Industrial chemical
FLAGYL	1-(2-hydroxyethyl)-2-methyl-5-nitroimidazole	Metronidazole	Protozoicide
FORMALDEHYDE	See Formalin, CH_2O	....	Preservative
FORMALIN (37–40%) commercial grade	37–40% formaldehyde in water	....	Preservative
FOSCHLOR	See Dylox	....	Insecticide
FRESCON	See n-trityl morpholine	....	Molluscicide
FUMAGILLIN	2,4,6,8-Decatetraenedioic acid 4- (1,2-epoxy-1,5-dimethyl-4-hexenyl)-5 methoxy-1-oxaspiro [2.5] oct-6-yl-ester	Amebacilin Fumidil	Antimicrobial
FURACIN, 4.59% water mix	See Nitrofurazone	....	Antibacterial
FURANACE	See P-7138	....	Antibacterial
FUROXONE	Furazolidone (N-5-nitro-2-furfurylidene)-3-amino-2-oxazolidone	Nf-180	Antibacterial
GAMMEXANE	See Benzene hexachloride	....	Insecticide
GENTIAN VIOLET	Methylrosaniline chloride	Bismuth violet Pyoktanin	Dye

[continued

Table 16. Chemical Names and Synonyms—*continued*

Name	Structure or Chemical Name	Synonyms	Common Use
GESAROL	See DDT	....	Insecticide
GLOBUCID	N^1-(5-ethyl-1,3,4-thiadiazol-2-yl) sulfanilamide	Albucidnatrium Sulfaethidole	Antimicrobial
GIX	1,1,1-Trichloro-2,2-bis-(p-fluorophenyl) ethane	DFD, DFDT	Insecticide
GRAMOXONE	See Paraquat	Paraquat	Herbicide
GRISEOFULVIN	7-chloro-4,6-dimethoxycoumaran-3-one-2-spiro-1′-(2′-methoxy-6′-methylcyclohex-2′-en-4′-one)	Fulcin Fulvicin	Antibiotic and antifungal
HALAMID	See Chloramine	....	Antiseptic
HYDROCHLORIC ACID	HCl	Muriatic acid	Industrial chemical
HYDROGEN PEROXIDE, 3% solution	H_2O_2	....	Antiseptic
IODOFORM	Triiodomethane	....	Antiseptic
IRON SULFATE	Did not specify if ferric or ferrous sulfate	....	Industrial chemical
KAMALA	Extract of glands and hairs on fruits of *Mallotus philippinensis*	....	Anthelmintic
KANAMYCIN	Derived from *Streptomyces kanamyceticus*	Cantrex Kantrex Kamycin Resistomycin	Antibiotic
KARATHANE	2,4-dinitro-6-(2-octyl) phenyl crotonate	Dikar Dinocap Mildex	Fungicide
KORLAN	O,O-dimethyl O-(2,4,5-trichlorophenyl)-phosphorothioate	Fenchlorophos Ronnel Trolene	Insecticide
LEXONE	See Benzene hexachloride	....	Insecticide

LILAC LEAVES	*Syringya vulgaris*	....	Garden shrub
LIME	$Ca(OH)_2 \cdot Mg(OH)_2$ or $Ca\ CO_3$	Caustic lime Hydrated lime Slaked lime	Fertilizer
LINDANE	Gamma isomer of benzene hexachloride	....	Insecticide
LYSOL	Commercial product of Lehn & Fink Division of Sterling Drug, Inc.	....	Disinfectant
MAGNESIUM SULFATE	$MgSO_4 \cdot 7H_2O$	Epsom salts	Cathartic
MALACHITE GREEN, zinc-free oxalate	p,p-benzylidenebis-N,N-dimethyl aniline	Aniline green Bright green Evergreen B Light green N Victoria green B or WB	Dye
MALATHION	0,0-dimethyl S-1,2, biscarbethoxyl ethyl phosphorothioate	....	Insecticide
MASOTEN	See Dylox	....	Insecticide
MENAZON	S-[(4,6-diamino-s-triazin-2)-yl] 0,0-dimethyl phosphorodithioate	....	Insecticide
MERCURIC CHLORIDE	$HgCl_2$	Corrosive sublimate	Industrial chemical
MERCURIC NITRATE	$Hg(NO_3)_2 \cdot H_2O$	Mercuric pernitrate	Industrial chemical
MERCUROCHROME	Disodium 2,7-dibrom-4-hydroxy-mercurifluorescein	Merbromin	Antiseptic
METASOL L	m-cresol-anylol	....	Disinfectant
METHYLENE BLUE	3,7-bis(dimethylamino)-phenazathionium chloride	Methylthionine chloride	Dye
METHYL PARATHION	0,0-Dimethyl 0-p-nitrophenyl phosphorothioate	Dalf Metacide Metron Nitrox 80	Insecticide
MICROPUR	German silver compound	....	Drinking water disinfectant

[continued

Table 16. Chemical Names and Synonyms—*continued*

Name	Structure or Chemical Name	Synonyms	Common Use
MITOX	p-Chlorobenzyl p-chlorophenyl sulfide	Chlorbenside	Miticide
NALED	1,2,-dibromo-2,2-dichloro-ethyl dimethyl phosphate	Bromex-50 Dibrom	Insecticide
NEGUVON	See Dylox	....	Insecticide
NICKEL SULFATE	$NiSO_4 \cdot 6H_2O$	....	Industrial chemical
NICLOSAMIDE	2′,5-Dichloro-4′-nitrosalicylanilide	Bayer 2353 Cestocid Fenasal Lintex Phenasal Yomesan	Anthelmintic
NICOTINE SULFATE	$(C_{10}H_{14}N_2)_2 \cdot H_2SO_4$	....	Insecticide
NIFURIPIRINOL	See P-7138	....	Antimicrobial
NITRIC ACID	HNO_3	....	Industrial chemical
NITROFURAZONE	5-nitro-2-furaldehyde semicarbazone	Furacin nfz	Bactericide
p-NITROPHENACYL CHLORIDE	Unknown	....	Molluscicide
NITROPHENYL AMIDINEURA	1-(p-nitrophenyl)-2-amidineura hydrochloride	T-72	Molluscicide
OSAROLUM	Unknown (Russian Origin)	....	Coccidiostat
OXALIC ACID	$(COOH)_2 \cdot n\ H_2O$	Ethanedioic acid	Industrial chemical
OXYTETRACYCLINE	4-(Dimethylamino)-1,4,4a,5,5a,6,11,12a-octahydro-3,5,6,10,12,12a-hexahydroxy-6-methyl-1,11-dioxo-2-naphthacenecarboxamide	Terramycin	Antibiotic
OZONE	O_3	....	Industrial chemical
P-7138	6-Hydroxymethyl-2,2(5 nitro-2 furyl) vinyl pyridine	Furanace Nifuripirinol	Antimicrobial

PAA-2056	Experimental compound (Parke-Davis)	....	Antiprotozoal
PARACHLOROPHENOXETHOL	Unknown (NIPA)	....	Bacteriostat
PARAFORMALDEHYDE	Polymerized formaldehyde	....	Disinfectant
PARAMOMYCIN	Paramomycin sulfate; derived from *Streptomyces rimosus*	Catenulin Humatin	Antibiotic
PARAQUAT, dichloride	1,1'-dimethyl-4,4'-dipyridinium dichloride	Bipyridilium Methyl viologen	Herbicide
PARATHION	See ethyl parathion	....	Insecticide
PENICILLIN, potassium	γ-chlorocrotylmercaptomethyl-penicillin potassium	Penicillin G	Antibiotic
PENTACHLOROPHENOL	C_6Cl_5OH	Dowicide G PCP Penta Santobrite Weed-beads Weedone	Wood preservative
PEROXYDOL	See Sodium perborate	....	Industrial chemical
PERU OIL	See Balsam of Peru Oil	....	Insecticide
PHENASAL	See Niclosamide	....	Anthelmintic
PHENOTHIAZINE	Thiodiphenylamine	Fenoverm Nemazine Phenovis	Anthelmintic
PHENOXETOL	1-Hydroxy-2-phenoxyethane	2-Phenoxyethanol	Disinfectant
PHILIXAN	Unknown	....	Unknown
PICRIC ACID	2,4,6-Trinitrophenol	Trinitrophenol	Industrial chemical
PINE NEEDLES	*Pinus* sp.	....	None
PLASMOCHIN	See Plasmoquin	....	Antimalarial

[continued

Table 16. Chemical Names and Synonyms—*continued*

Name	Structure or Chemical Name	Synoynms	Common Use
PLASMOQUIN	6-methoxy-8-(1-methyl-4-diethylamino) butylaminoquinoline salt of 2,2-dinaphthyl-methane 3,3-dicarboxylic acid	Pamaquine naphthoate Plasmochin naphthoate	Antimalarial
PMA	Pyridyl mercuric acetate	Phenylmercuric acetate TAG	Industrial chemical
POTASSIUM ANTIMONY TARTRATE	$C_4H_4KO_7Sb$	Tartar emetic	To treat schistosomiasis
POTASSIUM CHLORATE	$KClO_3$	....	Explosives
POTASSIUM CHLORIDE	KCl	....	Industrial chemical
POTASSIUM DICHROMATE	$K_2Cr_2O_7$	....	Industrial chemical
POTASSIUM HYDROXIDE	KOH	Caustic potash	Industrial chemical
POTASSIUM PERMANGANATE	$KMnO_4$	....	Industrial chemical
PR-3714	Experimental compound (Abbott)	....	Unknown
PRIASOL	German trade name	....	Unknown
PYRETHRUM	Extract of *Chrysanthemum* sp.	....	Insecticide
PYRETHRINS	See Pyrethrum	....	Insecticide
QUICKLIME	CaO	Hot lime	Industrial chemical
QUININE HYDROCHLORIDE	Extract of *Cinchona* sp. bark	....	Antimalarial
QUININE SULFATE	Extract of *Cinchona* sp. bark	....	Antimalarial
RIVANOL	6,9-diamino-2-ethoxyacridine lactate (Monohydrate)	....	Disinfectant
ROCCAL	Benzalkonium chloride; alkyl dimethyl benzyl ammonium chloride	Zephiran	Disinfectant
RONNEL	O,O-dimethyl O-(2,4,5 trichlorophenyl) phosphorothioate	Fenchlorophos Korlan Trolene	Insecticide
RUELENE	4-*tert*-Butyl-2-chlorophenyl N-methyl O-methylphosphoramidate	....	Oral insecticide
SALICYLIC ACID	Orthohydroxybenzoic acid	....	Industrial chemical

SALT	See sodium chloride, also sea water	....	Industrial chemical
SANTONIN	Extract of dried *Artemisia maritima*	Artemisin	Anthelmintic
SAPONIN	Sapogenin glycosides	....	Fish toxicant
SEA SALTS	Sea water	Instant Ocean	Aquariums
SILVER NITRATE	$AgNO_3$	....	Industrial chemical
SILVER PROTEIN	See Silvol	....	Disinfectant
SILVOL	Commercial formulation (Parke-Davis)	Silver protein (mild)	Disinfectant
SODIUM BORATE	$Na_2B_4O_7$	Sodium pyroborate Sodium tetraborate	Industrial chemical
SODIUM CHLORATE	$NaClO_3$	....	Industrial chemical
SODIUM CHLORIDE	NaCl	Rock salt Table salt	Preservative
SODIUM CHLORITE	$NaClO_2$	....	Industrial chemical
SODIUM FLUOSILICATE	Sodium hexafluorosilicate (Na_2Sif_6)	Salufer	Rodenticide
SODIUM HYDROXIDE	NaOH	Caustic soda	Industrial chemical
SODIUM HYPOCHLORITE	NaOCl	Household bleach Weed beads	Disinfectant
SODIUM PENTACHLOROPHENATE	See Pentachlorophenol	....	Wood preservative
SODIUM PENTACHLOROPHENOL	See Pentachlorophenol	....	Wood preservative
SODIUM PERBORATE PEROXYDOL	Unknown	?	Industrial bleach
SODIUM PEROXIDE PYROPHOSPHATE	Unknown	?	Unknown
SODIUM PYROPHOSPHATE PEROXYHYDRATE	Unknown	?	Disinfectant
STOVARSOL	See Acetarsone	....	Protozoicide

[continued

Table 16. Chemical Names and Synonyms—*continued*

Name	Structure or Chemical Name	Synonyms	Common Use
SULFAMETHAZINE, sodium	N'-(4,6-Dimethyl-2-pyrimidinyl) sulfanilamide	Diazil Mefenal Sulfamezathine Sulfamidine	Antimicrobial
SULFURIC ACID	H_2SO_4	....	Industrial chemical
SULQUIN	2-sulfanilamidoquinoxaline	Sulfaquinoxaline	Coccidiostat
TAG	Phenylmercuric acetate	PMA	Fungicide
TARTARIC ACID	$C_4H_6O_6$ (isomer unknown)	....	Industrial chemical
TEASEED CAKE	Pressed seeds of *Thea sinensis*	....	Insecticide
TERRAMYCIN	See Oxytetracycline	....	Antibiotic
TFM	See Trifluoromethyl nitrophenol	....	Lampricide
THIRAM	Bis(dimethylthiocarbamoyl) disulfide	Arasan Mercuram Thylate	Fungicide
TIGUVON	See Baytex	....	Insecticide
TINOSTAT	See dibutyltin dilaurate	....	Anthelmintic
TRIFLUOROMETHYL NITROPHENOL	3-Trifluoromethyl-4-nitrophenol	TFM	Lampricide
TRIS BUFFER	Commercial formulation containing Tris(hydroxymethyl)aminomethane	....	Physiological buffering solution
n-TRITYLMORPHOLINE	n-Tritylmorpholine	Frescon WL 8008	Molluscicide
TRYPAFLAVINE	Acriflavine hydrochloride	Acriflavine	Dye
TV-1096	Experimental compound (Parke-Davis)	....	Antiprotozoal
VIOLET K	Unknown	....	Russian dye
ZECTRAN	Methyl-4-dimethylamino-3,5-xylyl carbamate	....	Insecticide-molluscicide

ZEPHIRAN	Benzalkonium chloride	Roccal Zephirol	Disinfectant
ZINC CHLORIDE	$ZnCl_2$	Butter of zinc	Disinfectant
ZINC DIMETHYLDITHIOCARBAMATE	As given	Fuklasin Milbam Zerlate Ziram	Fungicide
ZIRAM	See Zinc dimethyldithiocarbamate	....	Fungicide
ZONITE	Stabilized sodium hypochlorite	....	Disinfectant

LITERATURE CITED

Agapova, A. I. 1957. Results of the study of fish parasites in waters of Kazakhstan. Trudi Inst. Zool. Akad. Nauk Kazakhskoi SSR 7. pp. 121–130.

—— 1966. Fish diseases and measures for combating them. Acad. Nauk Kazakstan SSR. Alma-Ata, 151 p. (In Russian.)

Alikunhi, K. H. 1957. Fish culture in India. Farm Bull. No. 20; Indian Counc. Agric. Res. 29: 144 pp.

Allen, K. O. and J. W. Avault, Jr. 1970. Effects of brackish water on Ichthyophthiriasis. Progr. Fish-Cult. 32(4): 227–230.

Allison, L. N. 1950. Common diseases of fish in Michigan. Mich. Dept. Cons. Misc. Publ. No. 5, 27 p.

—— 1954. Advancements in prevention and treatment of parasitic diseases of fish. Trans. Amer. Fish. Soc. 83 (1953): 221–228.

Allison, R. 1957a. Some new results in the treatment of ponds to control some external parasites of fish. Progr. Fish-Cult. 19: 58–63.

—— 1957b. A preliminary note on the use of di-n-butyl tin oxide to remove tapeworms from fish. Progr. Fish-Cult. 19: 128–130 and 192.

—— 1962. The effects of formalin and other parasiticides upon oxygen concentrations in ponds. Proc. 16th Ann. Conf. S.E. Assoc. Game and Fish Comm.: 446–449.

—— 1963. Parasite epidemics affecting channel catfish. Proc. 17th Ann. Conf. Southeastern Assoc. Game and Fish Comm. 346–347. (mimeo).: 1–3.

—— 1966. New control methods for *Ichthyopthirius* in ponds. FAO World Symp. on Warmwater Pond Fish Cult., FR: IX/E-9.

—— 1969. Parasiticidal activity of organophosphate compounds. Final report on project Al-00593. Dept. Zoology-Entomology (Fisheries). Auburn University, Auburn, Ala., 44 pp.

Amend, D. F. 1969. Progress in Sport Fishery Research. U.S. Bur. Sport Fish. Wildl., Resource Publication 88, Washington, D.C., p. 74.

Amlacher, E. 1961a. Die Wirking des Malachitgrüns auf Fische, Fischparasiten *(Ichthyophthirius, Trichodina)* Kleinkrebse und Wasserpflanzen. Deutsche Fisch. Zeit, 8(1): 12–15.

—— 1961b. Textbook of fish diseases. T.F.H. Publ., Neptune, N.J., 302 pp. (English translation.)

Anonymous. 1960. Hatchery Biologists Quarterly Report 1960, Region 1; Bur. Sport Fish. Wildl., Div. Fish Hatch., U.S. Dept. Interior, Washington, D.C.

—— 1968. Hatchery Biologists Quarterly Report 1968, Bur. Sport Fish. Wildl., Div. Fish Hatch., U.S. Dept. Interior, Washington, D.C.

—— 1969. Hatchery Biologists Quarterly Report, First Quarter 1969, Bur. Sport Fish. Wildl., Div. Fish Hatcheries, U.S. Dept. of the Interior, Washington, D.C. p. 18–19.

Applegate, V. C. and E. L. King. 1962. Comparative toxicity of 3-trifluormethyl-4-nitrophenol (TFM) to larval lampreys and eleven species of fishes. Trans. Amer. Fisheries Soc. 91(4): 342–345.

Arasaki, S. K., K. Nozawa, and M. Mizaki. 1958. On the pathogenicity of water mold. II. Bull. Japanese Soc. Sci. Fish. 23(9): 593–598.

Askerov, T. A. 1968. A method for control of saprolegnial fungus. Rybnoe Khozyaistvo, October 10: 23–24. (English translation—Bur. Sport Fish. Wildl., Washington, D.C.)

Astakhova, T. V. and K. V. Martino. 1968. Measures for the control of fungus diseases of the eggs of sturgeons in fish hatcheries. Problems of Ichthyology 8(2): 261–268. (English translation of Voprosy Ikhtiologiii by the American Fisheries Society.)

Avault, J. W. and R. Allison. 1965. Experimental biological control of a trematode parasite of bluegill. Exp. Parasitol. 17: 296–301.

Avdosev, B. S. 1962. New methods of malachite green used to control carp *Ichthyophthirius*. Rybnoe Khozyaistva. 38(7): 27–29. (In Russian.)

Avdosev, B. A. and N. E. Voznyi. 1963. Elimination of dactylogyrosis from rearing ponds. (Likvidatsiya daktilogiroza v vyrastnom prudu.) Vet. 40(8): 55–56. (Biol. Abstr. 45(16): 67492, 1964).

—— *et al.* 1962. The treatment and prophylaxis of pike infested by leeches. (Lechenie I Mery Profilaktiki Porazheniya Shchuk Pilyavkame.) Vet. 60: Biol. Abstr. 42: Abst. 8768.

Babaev, B. and A. Shcherbakova. 1963. The control *Bothriocephalus gowkongensis* from *Ctenopharyngodon idella.* Izvest. Akad. Nauk Turkmensk. SSR., Ser. Biol. Nauk. No. 4: 86–87. (In Russian.)

Bailosoff, D. 1963. Neguvon- ein wirsames Mittel zur Bekämpfung der Karpfenlaus und sonstiger parasitärer Fischkrankheiten. Deut. Fish. Zeit 10: 181.

Barbosa, F. S. 1961. Insoluble or slightly soluble chemicals as molluscicides. Bull. World Health Org. 25: 710–711.

Barney, E. 1963. Personal Communication. Spring Creek National Fish Hatchery, Underwood, Washington.

Batte, E. G., J. B. Murphy, and L. E. Swanson. 1951. New molluscicides for the control of freshwater snails. Amer. Jour. Vet. Research Vol. 12: 158–160.

Bauer, O. N. 1958. Parasitic diseases of cultured fishes and methods of their prevention and treatment. *In Parasitology of Fishes*, Edited by Dogiel, Petrushevski, and Polyanski, pp. 265–298. Oliver and Boyd, London, 1961, English translation; T.F.H., Neptune, N.J., 1970, reprint.

—— 1959. Parasites of freshwater fish and the biological basis for their control. Bull. State Sci. Res. Inst. Lake and River Fish. 49: 236 p. English translation OTS 61-31056, Dept. of Commerce, Washington, D.C.

—— 1966. Control of carp diseases in the USSR. FAO World Symp. on Warm-Water Pond Fish Cult. FR: IX/E-1: 344–352.

—— and B. Babaev. 1964. *Sinergasilus major* (Markevich, 1940), its biology and its pathological importance. Izv. Akad. Nauk Turkmen. SSR (Ser. Biol. Nauk), 3: 63–67.
—— and Yu. A. Strelkov. 1959. Diseases of artificially reared *Salmo salar* fry. Proc. 9th Conf. Fish Diseases, Acad. Sci. USSR. (Transl. OTS 61–31058, U.S. Dept. Commerce, pp. 89–93.)
—— and A. V. Upenskaya. 1959. New curative methods in the control of fish diseases. Proc. Conf. Fish Diseases, Acad. Sci. USSR, Ichthyological Committee. (Translated for the National Sci. Found. by the Israel Program Scientific Translations, Jerusalem), pp. 19–25.
Becker, C. D., and W. D. Brunson. 1968. The bass tapeworm: a problem in Northwest trout management. Progr. Fish-Cult. 30(2): 76–83.
Beckert, H. and R. Allison. 1964. Some host responses of white catfish to *Ichthyophthirius multifiliis*, Fouquet. 18th Ann. Meet. Southeastern Assoc. Game and Fish Comm. (Mimeo).
Bedell, G. W. 1971. Eradicating *Ceratomyxa shasta* from infected water by chlorination and ultraviolet irradiation. Progr. Fish-Cult. 33(1): 51–54.
Benoit, R. F. and N. A. Matlin. 1966. Control of *Saprolegnia* on eggs of rainbow trout (*Salmo gairdneri*) with ozone. Trans. Amer. Fish. Soc. 95(4): 430–432.
Bent, K. J. 1969. Fungicides in perspective. Endeavour (England) 28 (105): 129–134.
Bere, R. 1935. Further notes on the occurrence of parasitic copepods on fish of the Trout Lake region, with a description of the male of *Argulus biramosus*. Trans. Wisc. Acad. Sci. Arts & Letters 29: 83–88.
Berg, G. L. 1970. Farm Chemicals Handbook. Meister Publishing Co., Willoughby, Ohio, 469 pp.
Berrios-Duran, L. A., L. S. Ritchie, L. P. Frick and I. Fox. 1964. Comparative piscicidal activity of "stabilized Chevreul salt" (SCS), a candidate molluscicide and Bayluscide. Personal communication.
Bishop, H. 1963. Personal communication. National Fish Hatchery, Austin, Texas.
Bogdanova, E. A. 1962. Malachite green and formalin—effective agents for the control of trichodiniasis. Rybnoe Kohzyaistva. 38 (8): 30–31 (Biol. Abstr. 41: No. 17281). Eng. transl. FR, USFWS, Washington, D.C.
Borshosh, A. V. and V. V. Illesh. 1962. Elimination of *Ichthyophthirius* from fish ponds. Vet. 39, No. 11 (SLA trans. TT 66011296).
Boyce, C., T. W. T. Jones and W. A. Van Tongeren. 1967. The molluscicidal activity of N-Tritylmorpholine. Bull. World Health Org. 37: 1–11.
Bowen, J. T. and R. E. Putz. 1966. Parasites of Freshwater Fish; IV, Miscellaneous 3. Parasitic Copepod. *Argulus*. Fish Disease Leaflet No. 3, U.S. Dept. Interior, Bur. Sport Fish. Wildl., 4 pp.
Bradford, A. 1966. Personal communication. Benner Spring Fish Research Station, Belefonte, Pa.

Braker, W. P. 1961. Controlling salt water parasites. The Aquarium 30 (1): 12–15.
Brunner, G. 1943. Zur Bekampfung der Karpfenlaus *(Argulus foliaceus)* Fisch. Zeitschr. 46: 174–175.
Bureau of Sport Fisheries & Wildlife. 1968. Division of Fish Hatcheries, Hatchery Biologists Quarterly Report, Fourth Quarter—1968, p. 16.
—— 1969. Division of Fish Hatcheries, Hatchery Biologists Quarterly Report, First Quarter, 1969, p. 19.
Burrows, R. E. 1949. Prophylactic treatment for control of fungus *(Saprolegnia parasitica)* on salmon eggs. Progr. Fish-Cult. 11 (2): 97–103.
—— 1971. Personal communication. Fisheries Consultant, 2842 Magnolia St., Longview, Wash. 98632.
—— and B. D. Combs. 1968. Controlled environments for salmon propagation. Progr. Fish-Cult. 30 (3): 123–136.
—— and D. D. Palmer. 1949. Pyridylmercuric acetate: its toxicity to fish, efficacy in disease control, and applicability to a simplified treatment technique. Progr. Fish-Cult. 11 (3): 147–151.
Butcher, A. D. 1947. Ichthyophthiriasis in Australian trout hatchery. Progr. Fish-Cult. 9 (1): 21–26.
Bykhovskaya-Pavlovskaya, I. E., A. V. Gusev, M. N. Dubina, N. A. Izyumova, T. S. Smirnova, I. L. Sokolovskaya, G. A. Shtein, S. S. Shulman, and V. M. Epshtein. 1962. Key to the Parasites of Freshwater Fish of the U.S.S.R. Izdatel'stvo Akademii Nauk SSR, Moskova-Leningrad, 1962. Translation available from the Israel Program for Scientific Translation, Jerusalem, 1964, 919 pp.

Camey, T., E. Paulini and C. P. de Souza. 1966. Acao moluscicida do Gramoxone (N,N′-dimetil-p,p′-dipiridila) sobre *B. glabrata* em suas diversas fases de evolucao. Revista Brasileira de Malariologia e Doencas Tropicais. 18 (2): 235–245.
Campbell, A. S. 1950. Mail bag. Aquarium Jour. 21: 194–195.
Camper, J. 1970. Personal communication, National Fish Hatchery, Pisgah Forest, North Carolina.
Carothers, J. L. and R. A. Allison. 1966. Control of snails by the redear (shellcracker) sunfish. FAO World Symp. Warm-water Pond Fish Cult. FR: IX/E-11, 8 p.
Chabaud, A., R. Deschiens and Y. Le Corroller. 1965. Demonstration a marrakch d'un traitment molluscicide des eaux douces pa le chlorure cuivreux dans le cadre de la prophylaxie des bilharzioses. Bull. Soc. Path. Exot. 58 (5): 885–890.
Chang, S. 1960. Bull. Inst. Chem. Acad. Sinica No. 3: 44–50.
Chechina, A. S. 1959. Sanguinicolosis and measures for its control in the pond fisheries of the Belorussian SSR. Proc. 9th Conf. Fish Diseases, Acad. Sc. USSR (Transl. OTS 61-31058, U.S. Dept. of Commerce, 56–59).

Chemagro Corporation. 1971. Personal communication, P.O. Box 4913, Hawthorn Road, Kansas City, Mo.

Chen, Tung-Pai. 1933. A study on the methods of prevention and treatment of fish lice in pond culture. Lingnan Sci. J. 12 (2): 241–244.

Clemens, H. P. and K. E. Sneed. 1958. The chemical control of some diseases and parasites of channel catfish. Progr. Fish-Cult. 20 (1): 8–15.

—— 1959. Lethal doses of several commercial chemicals for fingerling channel catfish. USFWS Spec. Sci. Report-Fisheries No. 316, 10 p.

Combs, B. D. 1968. An electrical grid for controlling trematode cercariae in hatchery water supplies. Progr. Fish-Cult. 30 (2): 67–75.

Crandall, Catherine and C. J. Goodnight. 1959. The effect of various factors on the toxicity of sodium pentachlorophenate to fish. Limnol. and Oceanogr. 4 (1): 53–56.

Cross, D. G. 1971. Personal communication, Ministry of Agriculture Fisheries, and Food, Whitehall Place, London.

Crossland, N. O. 1967. Field trials to evaluate the effectiveness of the molluscicide N-tritylmorpholine in irrigation systems. Bull. World Health Organization 37: 23–42.

—— A. J. Pearson and M. S. Bennett. 1971. A field trial with the molluscicide Frescon for control of *Lymnaea peregra* Miller, snail host of *Diplostomum spathaceum* (Rudolphi). J. Fish. Biol. 3 (3): 297–302.

Cummins, R., Jr. 1954. Malachite green oxalate used to control fungus on yellow pikeperch eggs in jar hatchery operations. Progr. Fish-Cult. 16 (2): 79–82.

Davis, H. S. 1953. Culture and diseases of game fishes. University of California Press, Berkeley, 332 pp.

De, K. C. 1910. Report on the fisheries of Eastern Bengal and Assam. Shillong, Eastern Bengal and Assam Secretariate, 33 pp.

De Graaf, F. 1959. On the use of Lindane in fresh and sea water. Bull. Aquatic Biol. 1 (6): 41–43.

—— 1962. A new parasite causing epidemic infection in captive coral-fishes. Bulletin de L' Institut Oceanographique, Numero Special V A: 93–96. (1960.)

Dempster, R. P. 1955. The use of copper sulfate as a cure for fish diseases caused by parasitic dinoflagellates of the genus *Oodinium*. Zoologica. 40 (12): 133–139.

—— 1970a. Brackish water as a cure for ichthyophthiriasis in trout. Drum and Croaker 70 (1): 17.

—— 1970b. Sodium chlorite for water clarity in the marine dolphin system. Drum and Croaker 11 (3): 5–6.

—— and W. H. Shipman. 1969. The use of copper sulfate as a medicament for aquarium fishes and as an algaecide in marine mammal water systems. Occ. Papers Calif. Acad. Sci. No. 71, 6 pp.

—— 1970. The use of hydrogen peroxide in the control of fish disease. Drum and Croaker 70 (1): 27–29.

Deschiens, R. 1961. Nots aux prospecteurs sur les applications molluscicides chimiques en prophylaxis de pa bilharziose sur le terrain. Bull. Soc. Pathol. Exotique 54 (2): 365–375.
—— and H. Floch. 1964. Controle de l'action des molluscicides selectifs sur la microfaune et sur la microflore des eaux douces. Bull. de la Soc. de Pathol. Exotique 57: 292–299. Helm Absts. 34: 1006.
—— and M. Tahiri. 1961. Action molluscicide selective du sulfate de cadmium. Bull. Soc. Pathol. Exotique 54 (5): 944–946.
—— F. Floch and Y Le Corroller. 1963. Actions molluscicide et piscicide du sel cuprosulfitique de Chevreul en prophylaxie des bilharzioses. Bull. Soc. Pathol. Exotique 56 (3): 438–442.
—— A. Gamet, H. Brottes and L. Mvogo. 1965. Application molluscicide sur le terrain, au Camerain, de l'oxyde cuivereux dans le cadre de la prophylaxie des bilharzioses. Bull. Soc. Pathol. Exotique 58 (3): 445–455.
Detwiler, S. R. and G. E. McKennon. 1929. Mercurochrome (di-brom-oxy-mercuri-fluorescein) as fungicidal agent in growth of amphibian embryos. Anatomical Record 41 (2): 205–211.
Deufel, J. 1964. Direkte und indirekte Bekämpfung von *Diplostomum volvens* in kleinen Gewassern mit Bayluscid. Der Fischwirt 12: 1–3.
—— 1970. Untersuchungen mit dem Desinfectionsmittel Halamid. Fischwirt 20 (5): 114–117.
Dexter, R. 1963. Personal communication. Craig Brook National Fish Hatchery, East Orland, Maine.
Dogiel, V. A., G. K. Petrushevski, and Y. I. Polyanski. 1958. Parasitology of Fishes. Leningrad University Press, Translation available from T.F.H. Publ., Neptune, N.J. 1971, 384 pp.

Earnest, R. D. 1971. The effect of Paraquat on fish in a Colorado farm pond. Progr. Fish-Cult. 33 (1): 27–31.
Earp, B. J. and R. L. Schwab. 1954. An infestation of leeches on salmon fry and eggs. Progr. Fish-Cult. 16 (3): 122–124.
Edminster, J. O. and J. W. Gray. 1948. Toxicity thresholds from three chlorides and three acids to the fry of whitefish *(Coregonus clupeaformis)* and yellow pickerel *(Stizostedion v. vitreum)*. Progr. Fish-Cult. 10 (2): 105–106.
Ehrenford, F. A. 1968. Personal communication. Dow Chemical Co., Zionsville, Indiana.
Embody, G. C. 1924. Notes on the control of *Gyrodactylus* on trout. Trans. Amer. Fish Soc. 54: 48–50.
—— 1928. In Mellen, Ida, The treatment of fish diseases. Zoopathologica (N.Y. Zool. Soc.) 2: 21–31.
Engashev, V. G. 1966. *Raphidascaris* infection in fish. Veterinariya 43 (2): 59–61. (Helm. Abst. 36 (1): 175.)
Ergens, R. 1962. Direct control measures for some ectoparasites of fish. Progr. Fish-Cult. 24 (3): 133–134.

Erickson, J. D. 1965. Report on the problem of *Ichthyosporidium* in rainbow trout. Progr. Fish-Cult. 27 (4): 179–184.

Fasten, N. 1912. The brook trout disease at Wild Rose and other hatcheries. Rep. Comm. Fish. (Wisconsin) for 1911–1912: 12–22.

Ferguson, F. F., C. S. Richards and J. R. Palmer. 1961. Control of *Australorbis glabratus* by acrolein in Puerto Rico. U.S. Public Health Rep. 76 (6): 461–468.

Fischthal, J. 1949. *Epistylis*, a peritrichous protozoan on hatchery brook trout. Progr. Fish-Cult. 11 (2): 122–124.

Fish, F. F. 1933. The chemical disinfection of trout ponds. Trans. Amer. Fish. Soc. 63: 158–162.

—— 1939. Simplified methods for the prolonged treatment of fish diseases. Trans. Amer. Fish Soc. (1938.) 68: 178–187.

—— 1940. Formalin for external protozoan parasites. Progr. Fish-Cult. 48: 1–10.

—— and R. Burrows. 1940. Experiments upon the control of trichodiniasis of salmonid fishes by the prolonged recirculation of formalin solutions. Trans Amer. Fish. Soc. (1939) 69: 94–100.

Fletcher, A. 1961. Anchor worm (aquarium pest). All Pets 32 (2): 27–28.

Floch, H., R. Deschiens and Y. Le Corroller. 1964. Sur l'action molluscicide elective de l'oxyde cuivreux du ciuvre metal et du chlorure cuivreux. Bull. Soc. Pathol. Exotique 57: 124–138. (Helm. Abst. 34: 1007).

Foster, F. J. and L. Woodbury. 1936. The use of malachite green as a fish fungicide and antiseptic. Progr. Fish-Cult. 18: 7–9.

Foster, R. F. and P. A. Olson. 1951. An incident of high mortality among large rainbow trout after treatment with pyridylmercuric acetate. Progr. Fish-Cult. 13 (3): 129–130.

Frear, D. E. H. 1961. Pesticide Index. College Science Publishers, State College, Pennsylvania, 193 pp.

Frick, L. P., L. S. Ritchie, I. Fox and Wilma Jiminez. 1964. Molluscicidal qualities of copper protoxide (Cu_2O) as revealed by tests on stages of *Australorbis glabratus*. Bull. World Health Organization 30: 295–298.

Funnikova, S. V. and M. I. Krivova. 1966. Action of Chlorophos on the lower crustaceans. Uchen Zap. Kazansk. Vet. Inst. 96: 228–233.

Gamet, A., H. Brottes and Mvogo. 1964. Premiers essasis de lutte contre les vecteurs des bilharzioses dans les etangs d'une Station de Pisciculture au Cameroun. Bull. Soc. Pathol. Exotique 57: 118–124.

Gardner, W. and E. I. Cooke. 1968. Chemical synonyms and trade names. Chemical Rubber Co., Cleveland, Ohio, 635 pp.

Garibaldi, L. 1971. Chlorine + 3. Drum and Croaker 12 (1): 15–19.

Gerard, J. P. and P. de Kinkelin. 1971. Traitement de l'Acanthocephalose de la truite arc-en-ciel. La Piscicult. Francaise No. 26: 22–27.

Ghadially, F. N. 1963. Treatments for white-spot disease (Part 1). The Aquarist 28: 98–100.
—— 1964. Treatments for white-spot disease (Part 2). The Aquarist 29: 116–118.

Ghittino, P. 1968. Systemic control of hexamitiasis in trout fingerlings. Riv. Italiana Piscicult. Ittiopatol. 3 (1): 8–10.
—— 1970. Present status of whirling disease in Italian trout farms. Riv. Italiana Piscicult. Ittiopatol. 4: 89–92.
—— and G. Arcarese. 1970. Argulosi e Lerneosi dei pesci trattate in extenso con Masoten Bayer. Riv. Italiana Piscicult. Ittiopatol. 5 (4): 93–96.
Giudice, J. J. 1950. Control of *Lernaea carassii* Tidd, parasitic copepod infesting goldfish in hatchery ponds, with related observations on crayfish and the "fish louse," *Argulus* sp. M.S. Thesis, 1950. Library, University of Missouri, Columbia.
Glagoleva, T. P. and E. M. Malikova. 1968. The effect of malachite green on the blood composition of young Baltic salmon. Rybnoe Khozyaistva. (Fisheries) 44 (5): 15–18.
Gnadeberg, W., 1948. Beiträge zur Biologie und Entwicklung des *Ergasilus sieboldi* v. Nordmann (Copepoda: Parasitica). Zeitschr. Parasitenk. 14 (1 and 2): 103–180.
Goncharov, G. D. 1966. Effect of Chloramine-B on ectoparasites. Trudy Biol. Inland Waters 10: 338–340. Russian (Eng. transl. FRBC Library Bull. 5 (2): 2).
Goodrich, B. F. Co. 1966. Personal communication. B. F. Goodrich Tire Company, 500 South Main St., Akron, Ohio.
Gopalakrishnan, V. 1963. Controlling pests and diseases of cultured fishes. Indian Livestk. 1 (1): 51–54.
—— 1964. Recent developments in the prevention and control of parasites of fishes cultured in Indian waters. Proc. Zool. Soc. India, 17 (1): 95–100.
—— 1966. Diseases and parasites of fishes in warm-water ponds in Asia and the Far East. FAO World Symp. Warm-Water Pond Fish Culture. FR: IX/R-4: 319–343.
Gottwald, M. 1961. Die Anwendung von Malachitgrün und Kochsalz beim Erbrüten und Hältern von Laichfischen in Polen. Deutsche Fischerei-Zeit. 8 (2): 48–52.
Gowanloch, J. N. 1927. Notes on the occurrence and control of the trematode *Gyrodactylus*, ectoparasite on *Fundulus*. Trans. Nova Scotian Inst. Sci. 16: 126–131.
Grabda, E. 1965. Current studies on the control of parasitic diseases in fishes. Wiadomosci Parazytol. 11: 323–329. (Polish, Eng. Summ.)
Grabda, J. and E. Grabda. 1968 (1966). An attempt to control dactylogyrosis of carp with Neguvon. FAO Fish. Rep. No. 44, Vol 5, IX/E-7: 377–379.
Grétillat, S. 1965. Prophylaxie de la dracunulose par destruction des *Cyclops* au moyen d'un dérivé organique de synthèse, le diméthyldithio-

carbamate de zinc ou zirame. Biol Med. 54 (5): 529–539 (Helm. Abstr. 36 (1): 518).

Guberlet, J. E., H. A. Hansen and J. A. Kavanagh. 1927. Studies on the control of *Gyrodactylus*. Publ. Fish. Univ. Washington, College of Fisheries 2: 17–29.

Haderlie, E. C. 1953. Parasites of the fresh-water fishes of northern California. University of California Press, Berkeley and Los Angeles. 439 pp.

Harris, E. J. 1960. Quantitative determination of copper in a natural receiving water with 2,2 biguinoline. New York Fish & Game Journal 7 (2): 149–155.

Havelka, J. and I. Petrovicky. 1967. Curing Ich *(Ichthyophthirius multifiliis)* with malachite green. Trop. Fish Hob., January, 11–19.

—— F. Volf and J. Tesarcik. 1965. Investigation of new endoparasiticides with special regard to *Cryptobia cyprini* (Plehn, 1903) (Syn.: *Trypanoplasma cyprini* Plehn, 1903). Prace VURH Vodnany 5: 68–87 (Czech., Eng. and Germ. Summ.)

Herbert, D. W. M. and H. T. Mann. 1958. The tolerance of some freshwater fish for sea-water. Salmon Trout Mag. No. 153: 99–101.

Hess, W. N. 1930. Control of external fluke parasites on fish. J. Parasitol. 16: 131–136.

Hickling, C. F. 1962. Fish culture. Faber and Faber, London, 295 p.

Hindle, E. 1949. Notes on the treatment of fish infected with *Argulus*. Proc. Zool. Soc. London. 119 (1): 79–81.

Hnath, John G. 1970. Di-n-butyl tin oxide as a vermifuge on *Eubothrium crassum* (Bloch, 1779) in rainbow trout. Progr. Fish-Cult. 32 (1): 47–50.

Hofer, B. 1904. Hanbuch der Fischkrankheiten. Verlag Allg. Fischerei-Zeitung, Munich, 359 pp.

—— 1928. Quoted in Mellen (1928), The treatment of fish diseases. Zoopathologica (N.Y.) 2 (1): 1–31.

Hoffman, G. L. 1967. Parasites of North American Freshwater Fishes. University of California Press, Berkeley, 486 pp.

—— 1969. Parasites of freshwater fish. I. Fungi. *(Saprolegnia* and relatives) of fish and fish eggs. USFWS, Fish Disease Leaflet No. 21, 6 pp.

—— 1970. Progr. in Sport Fish. Res. (annual report). U.S. Dept. of the interior, Bureau of Sport Fisheries and Wildlife, Division of Fishery Research.

—— 1971. The effects of certain parasites on North American freshwater fishes. Proc. Internat. Congr. Limnol., Leningrad, U.S.S.R., (In Press).

—— 1972. Annual Report, EFDL, FR, BSFW, U.S. Dept. Interior, Wash. D.C., 20 pp.

—— and G. L. Hoffman, Jr. 1972. Studies on the control of whirling disease *(Myxosoma cerebralis)*. I. The effects of chemicals on spores

in vitro, and of calcium oxide as a disinfectant in simulated ponds. J. Wildl. Disease. 8: 49–53.
—— and R. E. Putz. 1964. Studies on *Gyrodactylus macrochiri* n. sp. (Trematoda: Monogenea) from *Lepomis macrochirus*. Proc. Helm. Soc. of Wash. 31 (1): 76–82.
—— 1966. Personal communication. Eastern Fish Disease Laboratory, Leetown (Kearneysville), W. Va.
—— and C. J. Sindermann. 1962. Common parasites of fishes. Circular No. 144, U.S. Fish Wildl. Serv., Washington. 17 pp.
Højgaard, M. 1962. Experiences made in Danmarks Akvarium concerning the treatment of *Oodinium ocellatum*. Bull. Inst. Oceanogr. Monaco. Numero Special 1A, Vol. A: 77–79.
Hora, S. L. 1943. The fish louse *Argulus foliaceous* Linnaeus, causing heavy mortality among carp fisheries of Bengal. Proc. Indian Sci. Congr., 39: 66–67.
—— and T. V. R. Pillay. 1962. Handbook on fish culture in the Indo-Pacific region. FAO Fish. Biol. tech. Pap., (14): 1–204.
Hoshina, T. 1966. On monogenetic trematodes. Gyobyokenkyu (J. Fish Path.) 1 (1): 47–57. Eng. transl., Bureau of Commercial Fisheries, U.S. Department of Interior.
Hsu, Me-keng and Jung-feng Jen. 1955. A preliminary report on the chemical control of the parasitic copepod, *Sinergasilus yui* Acta Hydrobiologia Sinensis 1955 (2): 59–68.

Hublou, W. F. 1958. The use of malachite green to control *Trichodina*. Progr. Fish-Cult. 20 (3): 129-132.
Hugghins, E. J. 1959. Parasites of fishes in South Dakota. Bull 484, Agric. Exper. Sta., South Dakota State College, Brookings. 73 pp.
Hunter, G. W. III. 1942. Studies on the parasites of fresh-water fishes of Connecticut. Bull. 63. Connecticut Geolo. Nat. Hist. Surv. Hartford. pp. 228–288.

Ivasik, V. M., and I. M. Karpenko. 1965. The application of lime for combatting ichthyophthiriasis (Primenenie izvesti v bor'be s ikhtioftiriozom). Veterinariya; Ezhemeyachnyi Nauchno-proizvodstven-nyi Zhurnal Ministerstva Sel' skogo Khozyaistva SSSR No 6: 1 p. (English translation: Division of Fishery Research, Bureau of Sport Fisheries and Wildlife.)
—— and V. S. Sutyagin. 1967. How to make sanitary carp fishery enterprises. Rybnoe Khozyaistvo 7: 23 (Engl. Transl. UDC 639.331.7).
—— O. I. Stryzhak, and V. N. Turkevich. 1968. On diplostomosis in the trout. Rybnoe Khozyaistvo 11: 27–28 (English translation by U.S. Dept. of the Interior, Bureau of Sport Fisheries and Wildlife, Washington, D.C.).
—— and B. G. Svirepo. 1964. The latest therapeutic and prophylactic preparations used directly in ponds. Rybnoe Khozyaistva 40 (11): 19–20.

Jackson, H. 1962. Personal communication. National Fish Hatchery, Lake Mills, Wisconsin.

Jobin, W. R. and G. O. Unrau. 1967. Chemical control of *Australorbis glabratus*. Public Health Report 82 (1): 63–71.

Johnson, A. K. 1961. Ichthyophthiriasis in a recirculating closed-water hatchery. Progr. Fish-Cult. 23 (2): 79–82.

Johnson, D. W. 1968. Pesticides and fishes—a review of selected literature. Trans. Amer. Fish. Soc. 97 (4): 398–424.

Johnson, H. E. 1956. Treatment of trichodinid infections of chinook salmon fingerlings. Prog. Fish-Cult. 18 (2): 94.

—— C. D. Adams, and R. J. McElrath. 1955. A new method of treating salmon eggs and fry with malachite green. Progr. Fish-Cult. 17 (2): 76–78.

Kabata, Z. 1970. Diseases of Fishes. Book I. Crustacea as Enemies of Fish. T.F.H. Publications, Jersey City, New Jersey, 171 pp.

Kanaev, A. 1967. Advances in fish diseases research. Rybovodstvo i Rybolovstvo 4 (3): 3–4.

Kasahara, S. 1962. Studies on the biology of the parasitic copepod *Lernaea cyprinacea* Linnaeus and the methods for controlling this parasite in fish culture ponds. Contrib. Fish. Lab., Fac. Agr., Univ. Tokyo, No. 3: 103–196 (Eng. Synopsis).

—— 1967. On the sodium pyrophosphate peroxyhydrate treatment for ectoparasitic trematodes on the yellow tail. Fish Pathology 1 (2): 48–53. (Japanese.)

—— 1968. Some external treatments for external parasites of cultured fish. Agriculture and Horticulture 43 (8): 1235–1238. (Japanese.)

Katz, M. 1961. Advises on copepods. U.S. Trout News, July-August.

Kelley, W. H. 1962. Controlling *Argulus* on aquarium carp. N.Y. Fish and Game Journal 9 (2): 118–126.

Kemp, P. S. J. T. 1958. Trout in Southern Rhodesia. V. On the toxicity of copper sulfate to trout. Rhod. Agric. Journal 55: 637–640 (Sport Fish. Abst. 1961, 6 (2): 95).

Kemper, H. 1933. Versuche über die Wirkung von Pyrethrumblütenpulver auf der Tiere verschiendener Klassen mit besonderer. Berüchsichtigung der wasserbewohnen den Arten. Zugleich ein Beitrag zur Frage der Anwendbarkeit des Pulvers bei der Bekämpfung tierischer Schädlinge in Wasserversorgungs un Abwasserbeseitigungsaulagen. Zeitsch. Gsundhtstech. un Städtehygr. 25 (3): 149–164.

Kerr, K. B. 1969. Personal communication. Salsbury Laboratories, Charles City, Iowa.

Khan, H. 1944. Study in diseases of fish. Infestation of fish with leeches and fish lice. Proc. Indian Acad. Sci. (B) 19 (5): 171–175.

Kimura, S. 1967. Control of the fish louse, *Argulus japonicus* Thiele, with Dipterex. (In Japanese.) Aquiculture, Tokyo, 8 (3): 141–150.

Kincheloe, J. 1964. Personal communication. Leavenworth National Fish

Hatchery, Leavenworth, Washington.
Kingsbury, O. R. and G. C. Embody. 1932. The prevention and control of hatchery diseases by treating the water supply. New York State Conservation Department.
Kiselev, V. K. and I. V. Ivleva. 1950. Measures against *Argulus*. Fish. Ind., Moscow XII (In Russian). Quoted *in* Dogiel, Petrushevski, and Polyanski (1958) (loc. cit.).
—— 1953. Some details on the biology of *Argulus* and measures against it under conditions of pond fishes. Trudy. Nauchno-Issled. Inst. Prud. i Ozern. Rechn. Ryb. Khozaistva Ukranian S.S.R., IX.
Klenov, A. P. 1970. Testing of anthelminthics against bothriocephaliasis of white amurs. Veterinariya 7:71–72. (Russian).
Knittel, M. D. 1966. Topical application of malachite green for control of common fungus infections in adult spring chinook salmon. Progr. Fish-Cult. 28 (1): 51–53.
Kocytowski, Br and Antychowicz. 1964. Anatomo—and histopathological lesions in ichthyophthiriosis of carp in sick fish and in those treated with malachite green. Bull. Vet. Inst. Pulawy 8 (3/4): 136–145.
Kokhanskaya, Y. M. 1970. The use of ultraviolet radiation for the control of disease in eggs and fishes (the MBU-3 compact bactercidal plant). J. of Ichthyology 10 (3): 386–393. (English translation by the American Fisheries Society).
Kubu, F. 1962. Heilung der mit *Ichthyophthirius* befallen durch Malachitgrun. Deutsche Fischerei Zeitung 9: 290.
Kulakovskaya, and Musselius. 1962. Quoted *in* Ivasik, V. M. and B. G. Svirepo, 1964. Rybnoe Khozyaistva 40: 19–20.
Kulow, H. and R. Spangenberg. 1969. Eigenschaften und Bedeutung der Nitrofurane fur die Prophylaxie und Therapie von Fischkrankheiten. Deutsche Fischerei Zeitung 16 (12): 365–371.
Kumar, A. K. 1958. Control of *Gyrodactylus* sp. on goldfish. M.S. Thesis Abstr., Auburn University, Auburn, Alabama (Unpublished).

Lahav, M. and S. Sarig. 1967. *Ergasilus sieboldi*, Nordman infestation of grey mullet in Israel fish ponds. Bamidgeh 19 (4): 69–80.
—— and M. Shilo. 1962. Development of resistance to Lindane in *Argulus* populations of fish ponds. Bamidgeh, Bull. Fish Cult., Israel 14 (4): 67–76.
—— S. Sarig, and M. Shilo. 1964. The eradication of *Lernaea* in storage ponds of carps through destruction of the copepodidal stage by Dipterex. Bamidgeh 16 (3): 87–94.
—— 1966. Experiments in the use of Bromex-50 as a means of eradicating the ectoparasites of carp. Bamidgeh 18 (3/4): 57–66.
Laird, J. and G. C. Embody. 1931. Controlling the trout gill worm (*Discocotyle salmonis*, Schaffer). Trans. Amer. Fish. Soc. 61: 189–191.
Lapter, V. I. 1967. An experiment on the sterilization of water with ultra-violet rays. Trudy Vsesoyuznogo Nauchno-Issledovatelskogo Inst. Prud.

Rybnogo Khozyaistva 15: 284–286 (Russian). Eng. transl. BSFW, Wash., D.C.

Larsen, H. L. 1963, 1964. Personal communication. U.S. Department of Interior, Bureau of sport Fisheries and Wildlife, Division of Fish Hatcheries, Washington, D.C.

Lavrovskii, V. and A. Uspenskaya. 1959. An effective method of controlling dactylogyriasis. Rybovodstvo i Rybolovstvo 6: 31–32 (Biol. Abstr. 46 (19): 83061).

Lawler, G. H. 1959. Biology and control of the pike whitefish parasitic worm, *Triaenophorus crassus* in Canada. Fish. Res. Bd. Canada. Prog. Rept. Biol. Sta. and Tech. Unit, London, 1: 31-37.

Lawrence, J. M. 1956. Preliminary results on the use of potassium permanganate to counteract the effects of rotenone on fish. Progr. Fish-Cult. 18 (1): 15–21.

Leger, L. 1909. La costiase et son traitement chez les jeunes alevins de truite. Compt. Rende. Acad. Sci., Paris 148 (19): 1284–1286.

Leith, D. A. and K. D. Moore. 1967. Pelton Pilot Hatchery, 1967 Progress Report, November 1966 through October, 1967. Oregon Fish Comm. Res. Div., Portland, 50 pp.

Leitritz, E. 1960. Trout and salmon culture. State of California, State Dept. of Fish and Game, Fish Bulletin No. 107, 169 pp.

Leteux, F. and F. P. Meyer. 1972. Mixtures of malachite green and formalin for controlling *Ichthyophthirius* and other protozoan parasites of fish. Progr. Fish-Cult. 34 (1): 21–26.

Lewis, S. 1967. Prophylactic treatment of minnow hatchery ponds with paraformaldehyde to prevent epizootics of *Gyrodactylus*. Progr. Fish-Cult. 29 (3): 160–161.

Lewis, W. M. 1961. Benzene hexachloride *vs* Lindane in the control of the anchor worm. Prog. Fish-Cult. 23 (2): 69.

—— and S. Lewis. 1963. Control of epizootics of *Gyrodactylus elegans* in golden shiner populations. Trans. Amer. Fish Soc. 92 (1): 60–62.

—— and J. D. Parker. 1965. Paraformaldehyde for control of *Gyrodactylus* and *Dactylogyrus*. Proc. 19th Ann. Conf. S.E. Assoc. Game and Fish Comm.: 222–225.

—— and M. G. Ulrich. 1967. Chlorine as quick-dip treatment for the control of gyrodactylids on the golden shiner. Progr. Fish-Cult. 29 (4): 229–231.

Loader, J. D. 1963. The use of chemicals for the aquarist. The Aquarist, 28: 28–29.

Locke, D. 1963. Personal communication. Department of Inland Fisheries and Game, Augusta, Maine.

Lotan, R. 1960. Adaptability of *Tilapia nilotica* to various saline conditions. Bamidgeh 12 (4): 96–100.

Loyen, A. 1931. Über die Möglichkeit der Bekämpfung der Karpfenlaus *(Argulus)* durch Trockenlegen der befallen Teiche und durch Anwendung von Atzkalk. Zeitsch. Fisch. 29: 597–604.

Mackenthun, K. M. 1958. The chemical control of aquatic nuisances. Comm. on Water Poll., St. Bd. of Health, Madison, Wisconsin, 64 p.
Malacca Research Institute. 1963. Rept. Trop. Fish Cult. Res. Inst., Malacca, Malaysia. 1962: 18 p.
Maloy, C. 1966. Personal communication. Dept. of Interior, Bureau of Sport Fisheries and Wildlife, Washington, D.C.
Markevich, A. P. 1967. Principles and ways of complex study of a parasitological situation in respect to arrangement of mass sanitation measures. Parasitology 1: 5–12. Acad. Nauk SSSR (Russian, Eng. summ.).
Marking, L. L. 1966. Investigations in fish control, 10. Evaluation of p,p′-DDT as a reference toxicant in bioassays. Bureau of Sport Fisheries and Wildlife, Dept. of Interior, Washington, D.C., 10 p.
—— and J. W. Hogan. 1967. Investigations in fish control, 19. Toxicity of Bayer 73 to fish. Bureau of Sport Fisheries, Dept. of Interior, Washington, D.C., 13 p.
Martin, R. L. 1968. Comparison of effects of concentrations of malachite green and acriflavine on fungi associated with diseased fish. Progr. Fish-Cult. 30(3): 153–158.
McAnnally, R. D. and D. V. Moore. 1966. Predation by the leech *Helobdella punctatolineata* upon *Australorbis glabratus* under laboratory conditions. J. Parasitol. 52: 196–197.
McElwain, I. B. and G. Post. 1968. Efficacy of Cyzine for trout hexamitiasis. Progr. Fish-Cult. 30(2): 84–91.
McKee, J. E. and H. W. Wolf. 1963. Water Quality Criteria. Calif St. Water Qual. Contr. Bd., Publ. No. 3A, 548 pp.
McKernan, D. L. 1940. A treatment for tapeworms in trout. Progr. Fish-Cult. 50:33–35.
McNeil, P. L. 1961. The use of benzene hexachloride as a copepodicide and some observations on lernaean parasites in trout rearing units. Progr. Fish-Cult. 23 (3): 127–133.
Meehean, O. L. 1937. *Dactylogyrus* control in ponds. Progr. Fish-Cult. Mem. I—131, 32: 10–12.
Meinken, H. 1954. Aquar.-u Terr. Z. 7: 50–51 (Taken from Stammer, 1959; Beitrage zur Morphologie, Biologie un Bekampfung der Karpfenlause. Zeitschrift. Parasitenk. 19: 135–208).
Mellen, I. 1928. The treatment of fish diseases. Zoopathologica 2: 1–31.
Merck & Co. 1968. The Merck Index. Merck and Company, Rahway, New Jersey, 1713 pp.
Merriner, J. V. 1969. Constant-bath malachite green solution for incubating sunfish eggs. Progr. Fish-Cult. 31 (4): 223–225.
Meyer, F. P. 1966a. Parasites of freshwater fish. II. Protozoa. 3. *Ichthyophthirius multifiliis*. U.S. Dept. of Interior, Bureau of Sport Fisheries and Wildlife, Fish Disease Leaflet No. 2, 4 p.
—— 1966b. A review of the parasites and diseases of fish in warm-water ponds in North America. FAO World Symp. on Warm-water Pond

Fish Cult. FR: IX/R-3: 290–318.
—— 1966c. A new control for the anchor parasite, *Lernaea cyprinacea.* Progr. Fish-Cult. 28 (1): 33–39.
—— 1967. Chemical control of fish diseases in warm-water ponds. Proc. Fish Farming Conf., Feb. 1–2, 1967. Texas A & M University, Texas Agricultural Extension Service, College Station, pp. 35–39.
—— 1969a. A potential control for leeches. Progr. Fish-Cult. 31 (3): 160–163.
—— 1969b. *Lernaea* control studies. Progress in Sport Fishery Research in 1969, U.S. Dept. of Interior, Bureau of Sport Fisheries and Wildlife, Washington, D.C. pp. 160–161.
—— 1970. Dylox as a control for ectoparasites of fish. Proceedings, 22nd Ann. Conf., Southeastern Assoc. Game and Fish Commissioners, p. 392–396.
—— and J. Collar. 1964. Description and treatment of a *Pseudomonas* infection in white catfish. Jour. Appl Microbiol. 12 (3): 201–203.
Meyer, M. C. 1962. The larger animal parasites of the fresh-water fishes of Maine. State of Maine, Dept. of Inland Fish and Game, Fish. Res. and Managm. Div., Bull. No. 1, 92 p.
Mitchum, D. L. and T. D. Moore. 1966. Efficacy of Di-N-butyl tin oxide on an intestinal fluke, *Crepidostomum farionis* in golden trout. Progr. Fish-Cult. 31 (3): 143–148.
Moore, J. P. 1923. A method of combating blood-sucking leeches in bodies of water controlled by dams. Publication T-267a, U.S. Dept. of the Interior, Fish and Wildlife Service, Washington, D.C., 2 p.
Mount, D. I. 1965. Personal communication, Environmental Protection Agency, Water Research Center, 4676 Columbia, Cincinnati, Ohio.
—— and C. E. Stephan. 1967. A method for establishing acceptable toxicant limits for fish—Malathion and the butoxyethanol ester of 2,4-D. Trans. Amer. Fish. Soc. 96 (2): 185–193.
Moyle, J. B. 1949. The use of copper sulfate for algal control and its biological implications. Limnological Aspects of Water Supply and Waste Disposal. Amer. Assoc. Adv. Sci. 1949, p. 79–87.
Musselius, V. A. 1967. Parasites and diseases of fish of phytophagous fishes and measures for combating them. Izdatel'stvo Kolos, Moskva, 1967. 82 p.
—— and N. T. Filippova. 1968. New preparations for combating ichthyophthiriasis in pond fish. Rybnoe Khozyaistvo Vol. 44 (1): 19–20.
—— 1969. Test of new preparations in control of ichthyophthiriasis in fish farms. Voprosy prudovro rybovodstva, Tome 16: 288–301 (Russian, Eng. summary).
—— and V. I. Laptev. 1967. Experimental application of Chlorophos for mollusc control at pond farms. Trudy vesesoyuznogo Nauchno-Issledovatel'skogo Instituta Prudovogo Rybnogo Khozyaistva, Voprosy

Prudovogo Rybovodstva. Pishchevaya Promyshlennost, Moscow, 15: 294–298. (English translation by Bur. of Sport Fish. and Wildl. Washington.)
—— and J. A. Strelkov. 1968. Diseases and control measures for fishes of Far-East Complex in farms of the U.S.S.R. Third Sympos. Comm. Off. Internat. Epizoot. Etude Malad. Poissons, Bull. of Off. Internat. Epizoot. 69 (9–10): 1603–1611.
Muzykovskii, A. M. 1968. Un anthelminthique pour la Bothriocephalose des carpes: le n-/2′-chlor-4′-nitrophényl 5-chlorsalicylamide. Bull. Off. Int. Epiz. 69 (9–10): 1539–1540.
—— 1971. The testing of phenasal in bothriocephalosis in carp. Trudy Vsesoyuznogo Nauchno-Issledovatel'skogo Inst. Prudovogo Rybnogo Khozyaistva 18: 146–148. (Engl. transl. SFWFR-TR-73-02 USDI.)

Nakai, N. and E. Kokai. 1931. On the biological study of a parasitic copepod *Lernaea elegans* Leigh-sharpe, infesting Japanese fresh water fishes. J. Imper. Fish. Exp. Stat. 2 (1): 123–128.
Naumova, A. M. 1968. Parasites of carp in pond fisheries and the diseases they cause. Itogi Nauki, Sci. Biol. (Zooparazitologiya 1966). VINITI: Moscow, pp. 66–82.
—— and A. I. Kanaev. 1962. Experience in the treatment of carp infected with coccidiosis. Vopr. Ikhtiol. 2 (4): 749–751.
Nazarova, N. S., A. M. Musikovski, A. N. Sorokin, R. N. Marchenckio, and V. I. Sen. 1969. The use of Fenasal against bothriocephaliasis in carp. Veterinariya 6: 57–59 (Russian).
Nechaeva, N. I. 1959. Parasitic diseases of *Salmo trutta caspius* fry in hatcheries and measures of their control. Proc. 9th Conf. on Fish Diseases, Acad. Sci. USSR (Trans. OTS 61–31058, Dept. of Commerce, 94–96).
Nelson, E. C. 1941. Carbarsone treatments for *Octomitus*. Progr. Fish-Cult. 55: 1–5.
Nichols, M. S., T. Henkel, and D. McNail. 1946. Copper in lake muds from lakes of the Madison area. Trans. Wisc. Acad. Sci., Arts, & Letters, 38: 333–350.
Nigrelli, R. F. and G. D. Ruggieri. 1966. Enzootics in the New York Aquarium caused by *Cryptocaryon irritans* Brown, 1951 (= *Ichthyophthirius marinus* Sikama, 1961), a histophagous ciliate in the skin, eyes and gills of marine fishes. Zoologica (New York) 51 (3): 97–101.
Normandeau, D. A. 1968. Disinfection of fish hatchery water supplies by means of ultraviolet irradiation. Progress Report F-14-R-3 State of New Hampshire, 10 p.
Northcote, T. G. 1957. Common diseases and parasites of fresh-water fishes in British Columbia. British Columbia Game Comm., Management Publ. No. 6, 25 p.

O'Brien. 1928. Quoted by Mellen (1928) *In* "The Treatment of Fish Diseases", Zoopathologica 2: 1–31.

O'Donnell, D. J. 1941. A new method of combating fungus infections. Progr. Fish-Cult. 56: 18–20.

O'Donnell, D. J. 1947. The disinfection and maintenance of trout hatcheries for the control of disease with special reference to furunculosis. Trans. Amer. Fish Soc., (1944) 74: 26–34.

Ortho Division. 1969. Unpublished data. Chevron Chemical Company, Ortho Division, Atlanta, Georgia.

Osborn, P. 1966. Effective chemical control of some parasites of goldfish and other pondfish. 1966 Ann. Meeting of the Wildlife Disease Association, Unpublished; Osage Catfisheries, Osage Beach, Mo. (Mimeo.)

Pasovskii, N. I. 1953. Quoted in Kanaev, 1967. Advances in fish disease research. Rybovodstvo: Rybolovstvo 4 (3): 3–4.

Patterson, E. E. 1950. Effects of acriflavine on birth rate. The Aquarium J. 21: 36.

Paulini, E. 1965. Reports on molluscicide tests carried out in Belo Horizonte, Brazil, 1964 WHO, Bilharziasis Res., Mol/Inf/20.65, p III/ 1–23.

—— and T. Camey. 1965. Un novo tipo de moluscicida com acao sistemica. Revista bras. Malar. Doenc. Trop. 17 (4): 349–353. (Eng. summ. p. 352).

Pavlovskii, E. N. (ed). 1959. Proceedings of the Ninth Conference on Fish Diseases. Izdatel'stow. Acad. Nauk. SSSR, Moscow-Leningrad. (Eng. trans. OTS 61-31058, 236 p.)

Peterson, E. J., E. W. Steucke, Jr., and W. H. Lynch. 1966. Disease treatment at Gavins Point Aquarium. The Dorsal Fin 6 (1): 18–19.

Pfeiffer, 1952. *In* Dogiel et al. 1958. *Parasitology of fishes.* Leningrad Univ. Press. (Trans. Oliver and Boyd, Ltd., London, 1961, 384 p.)

Plate, G. 1970, Masoten für die Bekämpfung von Ektoparasiten bei Fischen. Arch. f. Fischereiwissenschaft. 21 (3): 258–267.

Plehn, M. 1924. Praktikum der Fischkrankheiten. E. Schweizerbarts'che Verlagshandlung, Stuttgart, Germany, 179 p.

—— 1954. Quoted in *Fischkrankheiten* by W. Schäperclaus, 1954. Akademie-Verlag, 708 pp.

Popov, A. T. and G. Y. Jankov. 1968. Information sur les maladies des poissons et l'organization de la lutte contre ces maladies en Bulgarie. Third Symp. Comm. Off. Internat. Epizoot. Etude Malad. Poissons. Bull. Off. Int. Epiz., 69 (9–10): 1571–1576.

Post, G. and M. M. Beck. 1966. Toxicity, tissue residue, and efficacy of Enheptin given orally to rainbow trout for hexamitiasis. Progr. Fish-Cult. 28 (2): 83–88.

Postema, J. L. 1956. Ichthyophthiriose. Tydschr. Diergeneesk. 81 (1): 519–524. English Summary only seen.

Prevost, G. 1934. A criticism of the use of potassium permanganate in fish culture. Trans. Amer. Fish. Soc. 64: 304–306.
Prost, M. and M. Studnicka. 1967. Badania nad zastosowaniem estrów organicznych kwasu fosforowego w zwalczaniu pasozytów zewnetrznych ryb hodowlanych. III. Zwalczanie inwazji pierwotniaków z rodzaju *Chilodonella, Ichthyophthirius* i *Trichodina.* Medycyna Weterynaryjna 23 (4): 201–203.
—— 1968. Investigations on the use of pyro-phosphate for control of ectoparasites of cultured fishes. IV. Therapeutic value of the Polish preparation "Foschlor". Odbitka z "Medcyny Weterynaryjnej" 24 (2): 97–101.
—— 1971. Badania nad wartosciq terapeutycznq niektórych preparatów przy ichtioftiriozie karpi. Medycyna Weterynaryjna 28 (2): 69–73. (Polish, Eng. Summ.).
Prowse, G. A. 1965. Annual Report of the Tropical Fish-Cultural Research Institute. Batu Berendam, Malacca, Malaysia.
Prytherch. 1928. Quoted by Mellen, I. 1928. The treatment of fish diseases. Zoopathologica (N.Y. Zool. Soc.) 11: 1–31.
Putz, R. E. and J. T. Bowen. 1964. Parasites of freshwater fishes. IV. Miscellaneous. The anchor worm *(Lernaea cyprinacea)* and related species. U.S. Bureau of Sport Fisheries and Wildlife, Fisheries Leaflet No. 575, 4 p. (Now available as F.D.L.-12, 1968.)

Quebec Game and Fisheries Department. 1948. Control of leeches. Sixth Annual Report, Biological Bureau, Quebec Game and Fisheries Department, Montreal, pp. 85–87.

Radke, M. G., L. S. Ritchie and F. F. Ferguson. 1961. Demonstrated control of *Australobis glabratus* by *Marisa cornuarietis* under field conditions in Puerto Rico. Am. J. Trop. Med. and Hyg. 10 (3): 370–373.
Rankin, I. M. 1952. Treating fish affected by gill flukes. Water Life (London) Dec. 297–298.
Reddecliff, J. M. 1958, 1961. Formalin as a fungicide in the jar method of egg incubation. Notes for fish culturists, Benner Spring Fish Research Station, Pennsylvania Fish Commission (Unpublished).
Riechelt, H. W., Jr. 1971. Use of copper sulfate at Millen aquarium. Drum and Croaker 12 (1): 26.
Reichenbach-Klinke, H. 1966. Diseases and injuries of fish. Gustav Fischer Verlag, Stuttgart, 389 p. (In German.)
—— and E. Elkan. 1965. The principal diseases of lower vertebrates. Academic Press, London and New York. 600 p.
Ritchie, L. S. 1969. Personal communication. Puerto Rico Nuclear Center, Caparra Heights Station, San Juan, Puerto Rico 00935.
Robertson, R. 1954. Malachite green used to prevent fungus on lake trout eggs. Progr. Fish-Cult. 16 (1): 38.

Rogers, W. A. 1966. The biology and control of the anchor worm, *Lernaea cyprinacea* L. FAO World Symp. on Warm-water Pond Fish. Cult. FR: IX/E-10: 393–398.

Roth, W. 1910. Das Formalin als Vertilgungsmittel fur Aussenschmarotjer. Deutsche Fischerei-Correspondenz 14: 7–9.

—— 1922. Die Krankheiten der Aquarienfische und ihre Bekämpfung. Quoted by Mellen (1928) *In* Handbücher für die Prakt. Nat. Wiss. Arbeit, Franckh. Stuttgart. (Original not seen.)

—— 1928. *In* Mellen, Ida 1928. The treatment of fish diseases. Zoopathologica (N.Y. Zoo. Soc.) 11: 1–31.

—— 1954. Quoted by Schäperclaus (1954) in "Fischkrankheiten", Akademie-Verlag, Berlin, 708 p.

Rucker, R. R. 1958. Some problems of private trout hatchery operators. Trans. Amer. Fish. Soc. 87 (1957): 374–379.

—— W. O. Taylor, and D. P. Toney. 1963. Formalin in the hatchery. Progr. Fish-Cult. 25 (4): 203–207.

—— and W. J. Whipple. 1951. Effect of bactericides on steelhead trout fry. Progr. Fish.-Cult. 13 (1): 43–44.

Rychlicki, Z. 1966. Eradication of *Ichthyophthirius multifiliis* in carp. FAO World Symp. on Warm-Water Pond Fish Cult. FR: IX/E-3: 361–364.

Saha, K. C. and D. P. Sen. 1955. Gammexane in the treatment of *Argulus* and fish leech infection in fish. Ann. Biochem. and Experiment. Med. 15: 71–72.

—— and S. K. Chakraborty 1959. Gammexane in the treatment of a new parasitic infection in fish. Sci. Cult. 25 (2): 159.

Sakowicz, S. and S. Gottwald. 1958. Zapobieganie i zwalczanie plesni u tarlaków troci i lososi przy pomocy kapieli w roztworze zieleni malachitowej. Roczniki Nauk Rolniczych 73B: 281–293.

Sanders, J. E., J L Fryer, D. A. Leith, K. D. Moore 1972. Control of the infectious protozoan *Ceratomyxa shasta* by treating hatchery water supplies. Prog. Fish-Cult. 34 (1): 13–17.

Sandler, A. 1966. Diseases of tropical fish, pp. 511–514 *in* Catcott, E. J. and J. F. Smithcors (ed) Progress in feline practice. Amer. Veterinary Publ. Wheaton, Ill. 649 pp.

Sarig, S. 1966. A review of diseases and parasites of fishes in warm-water ponds in the Near East and Africa. Proc. FAO World Symp. on Warm-water Pond Fish Cult. FR: IX/R-2: 278–289.

—— 1968. Possibilities of prophylaxis and control of ectoparasites under conditions of intensive warm-water pondfish culture in Israel. Bull. Off. Int. Epizoot., 69 (9–10): 1577–1590.

—— 1969. Pesticide concentrations for pond spraying effective against various ectoparasites attached to carp. (mimeo). 1 page. The Hebrew University-Hadassah Medical School, Jerusalem; Dept. of Microbiological Chemistry and the Field Laboratory at Nir David.

—— and M. Lahav. 1959. The treatment with lindane of carp in fish ponds infected with the fish-louse, *Argulus*. Proc. Tech. Papers Gen. Fish. Counc. Medit. 5: 151–156.
—— M. Lahav, and M. Shilo. 1965. Control of *Dactylogyrus vastator* on carp fingerlings with Dipterex. Bamidgeh 17 (2): 47–52.
Schäperclaus, W. 1931. XXI. Die Drehkrankheit in der Forellensucht und ihre Bekämpfung. Zeitschrift für Fischerei 29: 521–567. (English translation by Eastern Fish Disease Lab, Leetown, W. Va.)
—— 1932. Die Drehkrankheit und ihre Bekämpfung. Mitt d. Fischereivereine, Westausgabe, Vol. 2: 26–33.
—— 1954. Fischkrankheiten. Akademie-Verlag, Berlin, 708 p.
Schneberger, E. 1941. Fishery research in Wisconsin. Progr. Fish-Cult. No. 56: 14–17.

Schnick, R. A. 1972. A review of literature on TFM (3-trifluormethyl- 4-nitrophenol) as a lamprey larvicide. Investigations in Fish Control. 44. U.S. Bureau of Sport Fisheries and Wildlife, Washington, D.C., 31 p.
Scott, W. W. and C. O. Warren, Jr. 1964. Studies of the host range and chemical control of fungi associated with diseased tropical fish. Va. Agr. Expt. Sta. Tech Bull. 171: 1–24.
Seale. 1928. Quoted by Mellen, I. 1928. The treatment of fish diseases. Zoopathologica (N.Y.) 2: 1–31.
Shilo, M. S., S. Sarig, and R. Rosenberger. 1960. Ton scale treatment of *Lernaea* infected carps. Bamidgeh 12 (2): 37–42.
Shimizu, M. and Y. Takase. 1967. A potent chemotherapeutic agent against fish diseases: 6-hydroxymethyl-2-[2-(5-nitro-2-furyl) vinyl] pyridine (P-7138). Bull. Jap. Soc. Sci. Fish. 33 (6): 544–554.
Sindermann, C. J. 1953. Parasites of fishes of north central Massachusetts. Mass. Dept. of Conserv., Boston, 28 p.
Slater, D. C. 1952. New treatment for white spot. Water Life, June: 122.
Smith, W. W. 1942. Action of alkaline acriflavine solution on *Bacterium salmonicida* and trout eggs. Proc. Soc. Exper. Biol. and Med. 51: 324–326.
Smith, R. T. and E. Quistorff. 1940. The control of *Octomitus*: Calomel in the diet of hatchery salmon. Progr. Fish-Cult. 51: 24–26.
Snow, J. 1958 and 1962. Personal communication. National Fish Hatchery, Marion, Alabama.
Snow, J. R. and R. O. Jones. 1959. Some effects of lime applications to warm-water hatchery ponds. S. E. Assoc. Game and Fish Comm., 13th Ann. Conf.: 95–101.
Sokolov, P. M. and N. G. Maslyukova. 1971. The use of malachite green in gyrodactylosis of carp. Vsesoyuznogo Nauchno-Issledovatel'skogo Inst. Prudovoga Rybnogo Khozyaistva 18: 179–180. (Engl. Transl. SFWFR-TR-73-01, Div. Fish Res., U.S. D.I.).
Southwell, T. 1915. Notes from Bengal Fisheries Laboratory, Indian Museum. 2. On some Indian parasites of fish with a note on carcinoma in trout. Rec. Indian Mus. 11: 311–330.

Spall, R. D. 1970. Possible cases of cleaning symbiosis among freshwater fishes. Trans. Amer. Fish. Soc. 99 (3): 599–600.

Sproston, N. 1956. The effect of insecticide "666" on some fish pests and other animals in the fish ponds. Acta Hydrobiologica Sinica 1: 89–98.

Stammer, H. J. 1959. Beiträge zur Morphologie, Biologie, und Bekämpfung der Karpfenläuse. Zeitschr. Parasitenk., Berlin 19 (2): 135–208.

Steffens, W. 1962. Verhütung des Saprolegnia-Befalls von Forellenneiern durch Formalin. Deutsche Fisch. Zeit. 9: 287–289.

Stiles. 1928. Quoted by Mellen, I., 1928. The treatment of fish diseases. Zoopathologica (N.Y.) 11 (1): 1–31.

Stolk, A. 1956. Cited in *Diseases of Fishes* by C. Van Duijn, 1956, Water Life, Dorset House, London, pp. 29–30.

Sukhenko, G. E. 1963. *Argulus pellucidus* Wagler, 1935 (Crustacea, Brachiura), a species new to the fauna of the USSR, found in the ponds of the Ukraine. Zool. Zh. 42 (4): 621–622.

Surber, E. W. 1948. Chemical control agents and their effects on fish. Progr. Fish Cult. 10 (3): 125–131.

—— 1969. Personal communication. State Fish Hatchery, Front Royal, Virginia.

Swingle, H. S. 1955. Personal communication, Auburn University, Auburn, Alabama.

Tack, E. 1951. Bekämpfung der Drehkrankheit mit Kalkstickstoff. Der Fischwirt, No. 5: 123–129.

Takase, Y., K. Kouno, and M. Shimizu. 1971. Effect of Nifurpirinol against diseases caused by bacteria and protozoa in hobbyfishes. Fish Pathology (Japan) 5 (2): 81–84.

Tebo, L. B. and E. G. McCoy. 1964. Effect of sea-water concentration on the reproduction and survival of largemouth bass and bluegills. Progr. Fish-Cult. 26 (3): 99–106.

Tesarcik, J. and J. Havelka. 1966. Pruzkum antiparazitarnich a protiplisnovych opatreni. Prace VURH Vodnany 6: 97–122.

—— 1967. Prevention and cure of fish diseases. Ustav Vedeckotechnickych Informaci NZLH (In Czech).

—— and J. Mares. 1966. The applicability of an antiparasitic bath in malachite green of the sheatfish fry (*Siluris glanis* L.) Buletin VUR, Vodnany 2 (3): 13–15 (In Czech, Eng. summ. p. 15).

Tripathi, Y. R. 1954. Studies on parasites of Indian fishes. III. Protozoa. 2. (Mastigophora and Cliophora). Rec. Indian Mus. 52 (2–4): 221–230.

Van Cleave, H. J. and J. F. Mueller. 1934. Parasites of Oneida Lake Fishes. Part III. A biological and ecological survey of the worm parasites. Roosevelt Wildlife Annals. 3 (3–4): 161–334.

Van Duijn, C. 1956. Diseases of fishes. Water Life, Dorset House, London, 174 p.

—— 1967. Diseases of fishes. Iliffe Books Ltd., London, 309 p.
Van Roekel, H. 1929. Acetic acid as a control agent for *Cyclochaeta* and *Gyrodactylus* in hatchery trout. California Fish and Game. 15 (3): 230–233.
Venulet, J. and G. O. Schultz. 1964. A new molluscicide 1-(p-nitrophenyl)-2-amidineurea hydrochloride (T72). Nature 204 (4961): 900–901.
Vik, R. 1965. Studies of the helminth fauna of Norway. VI. An experiment in the control of *Diphyllobothrium* infections in trout. Medd Fra Zool. Mus., Oslo, Contr. No. 75:76–78. (In English.)
Villiers, J. P. de and J. C. Mackenzie. 1963. Structure and activity in molluscicides; the phenacyl halides, a group of potentially useful molluscicides. Bull. World Health Org. 29 (3): 424–427.
Vlasenko, M. I. 1969. Ultraviolet rays as a method for the control of fish eggs and young fishes. Prob. Ichthyology 9 (5): 697–705 (Engl. Translation—American Fish Soc.).

Watanabe, M. 1940. Salmon culture in Japan. Progr. Fish-Cult. No. 48: 14–18.
Wellborn, T. L., Jr. 1965, 1966, and 1967. Personal communication. Fish. Dept., Auburn University, Auburn, Alabama.
—— 1969. The toxicity of nine therapeutic and herbicidal compounds to striped bass. Progr. Fish-Cult. 31 (1): 27–32.
—— 1971. Toxicity of some compounds to striped bass fingerlings. Progr. Fish-Cult. 33 (1): 32–36.
Wellborn, T. L. and W. A. Rogers. 1966. A key to the common parasitic protozoans of North American fishes. Auburn University, Zool. Ent. Dept. Series, Fisheries No. 4, 17 p.
Willford, W. A. 1967a. Investigations in fish control, 18. Toxicity of 22 therapeutic compounds to six fishes. U.S. Dept. of the Interior, Bureau of Sport Fisheries and Wildlife, Washington, D.C. 10 p.
—— 1967b. Investigations in fish control, 20. Toxicity of dimethyl sulfoxide (DMSO) to fish. U.S. Dept. of Interior, Bureau of Sport Fisheries and Wildlife, Washington, D.C. 18 p.
Wilson, C. O. and T. E. Jones. 1962. American Drug Index. J. B. Lippincott, Philadelphia, 840 p.
Wolf, H. 1957. Quoted in Rucker, 1957. No publication.
Wolf, L. E. 1935a. The use of potassium permanganate in the control of fish parasites. Progr. Fish-Cult. No. 11: 20–21.
—— 1935b. The use of potassium permanganate in the control of fish parasites. Trans. Amer. Fish Soc. 65: 88–100.
Wood, J. 1968. Diseases of Pacific salmon; their prevention and treatment. Washington State Dept. of Fisheries. Olympia, Washington. 74 p.
Woyanarovich, E. 1954. Eine neue Methode zur Bekämpfung der Ektoparasiten von Karpfen. Acta Veterin. Acad. Sci. Hung. Band 4, No. 1.

—— 1959. Erbrütung von Fischeiern im sprühraum. Arch. Fischereiwissenschaft, 10 (3): 169–232.

Yasutake, W. T., D. R. Buhler, and W. E. Shanks. 1961. Chemotherapy of hexamitiasis in fish. J. Parasitol. 47 (1): 81–86.

Yin, W. Y., M. E. Ling, G. A. Hsu, I. S. Chen, P. R. Kuang, and S. L. Chu. 1963. Studies on the lernaeosis *(Lernaea,* Copepoda parasitica) of the freshwater fishes of China. Acta Hydrobiologica Sinica 1963 (2): 48–117.

Yousuf-Ali, M. 1968. Investigation on fish diseases and parasites in East Pakistan. Third Symp. Comm. Int. Epizoot. Etude Malad. Poissons, Stockholm, 1968. Contrib. No. 27, 5 pp.

Yui-fan, *et al.* 1961. Fishing industry of the inland waters of China. *Quoted in:* Parasites and diseases of fish of phytophagous fishes and measures for combating them by V. A. Musselius, 1967. Izdatel'stvo Kolos, Moskva, 82 pp. (Original not seen.)

Zeiller, W. 1966. Anthium dioxide, a new disinfectant compound for aquaria. Ichthyologica 37 (3): 107–110.

Zschiesche, A. 1910. Formalin, a new treatment for costiasis. Allgemeine Fischerei-Zeitung 35: 147–149. (German, article not seen, quoted by Schäperclaus, pers. comm., 1969).